SYMBOLIC CALCULUS FOR SEMILINEAR HYPERBOLIC PROGRESSING WAVES

SYMBOLIC CALCULUS SEMILINEAR HYPERBOLIC PROGRESSING WAVES

H. BOUGRINI AND A. PIRIOU
EDITORS

Nova Science Publishers, Inc.
Huntington, New York

Library of Congress Cataloging-in-Publication Data

Symbolic calculus for semilinear hyperbolic progressing waves/ H. Bougrini ... [et al.]
p. cm.
Includes bibliographical references and index.
ISBN 1-56072-878-8
1. Differential equations. Hyperbolic—Numerical solutions. 2. Singularities (Mathematics) 3. Nonlinear wave equations. I. Bougrini, H. (Hassane)

QA377.S962 2000
515'.35--dc21

227 Main Street, Suite 100
Huntington, New York 11743
Tele. 631-424-6682Fax 631-424-4666
e-mail: Novascience@earthlink.net
Web Site: http://www.nexusworld.com/nova

Printed in the United States of America

Contents

PREFACE

The main purpose of this book is to give a self contained synthesis of different results obtained in the last few years by H. BOUGRINI, G. LASCHON, B. NADIR, A. PIRIOU, J.P. VARENNE in the domain of symbolic calculus of conormal singularities of semilinear hyperbolic progressing waves. This study was led through diverse collaborations involving mainly the Laboratory of Mathematics of Nice Sophia-Antipolis University (France) and the Department of Mathematics and Computer Science of the Faculty of Sciences of Meknès (Morocco).
We are also grateful to the French-Moroccan programme -Action Intégrée - Harmonic Analysis and Partial Differential Equations coordinated by the Laboratory of Mathematics of Nice Sophia-Antipolis University and the Department of Mathematics and Computer Science of Faculty of Sciences of Ain-Chock, Casablanca, which made possible exchanges leading to the final redaction of this book.
It is assumed that the reader is familiar with the basic notions of the classical theory of pseudodifferential operators and linear microlocal analysis. The first two chapters are devoted to a complete exposition of J.M. Bony's basic results on propagation and interaction of conormal progressing waves, including the linear and nonlinear study of the appropriate conormal spaces. We have avoided a systematic use of the paradifferential calculus, referring only to some properties of the operator of multiplication by functions in Sobolev or Besov Spaces, and using simple linearization techniques adapted to the conormal context.

Hassane Bougrini and Alain Piriou, Editors

INTRODUCTION

Let $L = \sum_{j=1}^{n} A_j(x)D_{x_j} + A(x)$ be a $N \times N$ first order linear differential operator with smooth $N \times N$ matrix valued coefficients in $\Omega \subset I\!R^n$. The principal symbol l of L is defined by $l(x,\xi) = \sum_{j=1}^{n} \xi_j A_j(x)$ when $(x,\xi) \in \Omega \times (I\!R^n \setminus 0)$, and we set $p(x,\xi) = detl(x,\xi)$. Assuming that p is real valued and that the characteristic set $p^{-1}(\{0\})$ is non empty, we are interested in the propagation of the singularities of the solutions u of a semilinear system of the form $Lu = f(x,u)$, when f is a smooth vector valued function.

If f is linear with respect to u, a fundamental result of propagation of microlocal singularities has been given by L. Hörmander : the wave front set of u is invariant by the flow associated to the hamiltonian field of p. If the source term f is not linear with respect to u, simple examples show that this invariance is no longer true in general : purely non linear phenomena due to self spreading or crossing occur, which have been studied by many authors for more than twenty years, motivating new tools of nonlinear microlocal analysis such as paradifferential calculus and successive microlocalisations. More particularly, important results were obtained under conormality assumptions. Recall that if Σ is a smooth closed hypersurface of Ω, the space $H_s(\Sigma)$ of conormal distributions of index s with respect to Σ is constituted of all distributions in the Sobolev space $H_s^{loc}(\Omega)$ or in the Besov space ${}^\infty H_s^{loc}(\Omega)$ which remain in the same space under iterated action of smooth vector fields tangent to Σ. Typical examples are the multihomogeneous functions $a(x')\mathcal{H}(\pm x_1)|x_1|^\alpha(ln|x_1|)^k$ when $\Sigma = \{x = (x_1, x') \in I\!R \times I\!R^n / x_1 = 0\}$, $\alpha \in \mathbb{C}$, $\Re\alpha > 0$, or $\alpha = 0$, $k \in I\!N$, $a \in C^\infty$. The classical conormal functions with respect to Σ are asymptotic sums of multihomogeneous functions (locally), and include the piecewise smooth functions with respect to Σ, which correspond to asymptotic sums with $\alpha \in I\!N$, $k = 0$.

The wave front set of an element of $H_s(\Sigma)$ is included in the conormal bundle $N(\Sigma)$ to Σ. Moreover, when $s > 1/2$, the space $H_s(\Sigma)$ is invariant under the action of smooth functions, thus if $u \in H_s(\Sigma)$ satisfies $Lu = f(x,u)$ and is effectively singular on Σ, $N(\Sigma)$ must be contained in $CharL$, that is Σ must be characteristic for L. Such a solution u is called a (single) conormal progressing wave. Under strict hyperbolicity assumptions and considering solutions u in H_s^{loc} when s is large enough, J.M. Bony has proved the propagation of a single conormal wave. Moreover, if Σ_1, Σ_2 are two smooth closed characteristic hypersurfaces intersecting transversally in the future, and if $u \in H_s(\Sigma_1 \cup \Sigma_2)$

in the past, J.M. Bony has proved that u belongs to the conormal space $H_s(\Sigma_1 \cup \Sigma_2 \cup ... \cup \Sigma_r) = \sum_{j=1}^{r} H_s(\Sigma_j, \Gamma)$, where $(\Sigma_1, \Sigma_2, ..., \Sigma_r)$ is the bunch of characteristic hypersurfaces issued from $\Gamma = \Sigma_1 \cap \Sigma_2$, $H_s(\Sigma_j, \Gamma)$ being defined as before (the vector fields tangent to Σ are replaced by the vector fields tangent to Σ_j anf Γ). This proves that, after interaction at Γ of the two incident conormal progressing waves, the singularities of u are located on the forward characteristic hypersurfaces $\Sigma_1^+, \Sigma_2^+, ..\Sigma_r^+$ issued from Γ, and moreover u is a conormal progressing wave with fronts $\Sigma_1^+, \Sigma_2^+, ..\Sigma_r^+$. Thus, apart from the singularities on Σ_1^+ and Σ_2^+ which are already present in the linear case, singularities of "nonlinear type" are expected on Σ_j^+ $(3 \leq j \leq r)$. These singularities are proven to be weaker than the two incident singularities. But at this level several questions remained open: Are we sure that these nonlinear type singularities really appear? what kind of nonlinearity of the source term is requested to warrant this appearance? In this case, what are the optimal indexes, size and nature of the appearing conormal progressing waves? These questions have been solved in several steps by the authors of this book. The key is the consideration of the oscillatory aspect of conormal distributions. L. Hörmander has proven that conormal distribution of Besov index s are characterized by local Fourier integral expressions of the form $u(x) = \int e^{ix_1\xi_1} a(x', \xi_1) d\xi_1$ when $\Sigma = \{x = (x_1, x') \in I\!R \times I\!R^{n-1} / x_1 = 0\}$, where $a(x', \xi_1)$ is a symbol of degree $\mu = -s - 1/2$. This leads to the intrinsic definition of the principal symbol. Returning to our questions, it is natural to study the principal symbols of u (outside Γ) corresponding to significant indexes. Firstly, using refined conormal spaces $\sum_j I^{\mu_j, \nu_j}(\Sigma_j, \Gamma)$ allowing different indexes on $\Sigma_1^+, \Sigma_2^+, ..\Sigma_r^+$ and precising an additional conormal regularity with respect to Γ, we improve the conormality indexes of Bony's study. Then we show that these improved indexes are (in general) optimal: the corresponding principal symbols are determined from the past thanks to transport equations. More precisely, in the case of Σ_j^+ when $3 \leq j \leq r$, we have an initial condition at the wedge Γ in terms of a product of the two incoming principal symbols by a convenient second derivative of $f(x, u)$ with respect to u. Thus the role of the nonlinearity of the source term is clarified, and the optimal conormality of a non linear type singularity is given by a natural sum law if we consider the degrees of symbols rather than the Besov indexes.

Besides this general study of principal symbols, we show that a complete symbolic calculus is possible in the case of a single progressing wave for a general $N \times N$ system and in the case of transversal interaction for 2×2

systems, including properties of propagation and interaction of classical conormal and piecewise smooth progressing waves. We study also a phenomenom pointed out by G. Métivier and J. Rauch : when the number r of characteristic hypersurfaces issued from Γ is strictly smaller than the size N of the system, the incoming piecewise smooth regularity may be destroyed after transversal interaction. The obstruction to the conservation of piecewise smoothness don't appear at the principal symbol level. A precise microlocal analysis is performed in an example thanks to a study of complete symbols.

For sake of simplicity of presentation, we deal generally with real matrix valued coefficients and with real vector valued solutions, but the complex case is similar. We consider also $N \times N$ first order systems rather than high order scalar equations, because the polarization properties of symbols are less natural in the latter case. Moreover, although we assume generally that the real characteristics are simple, our methods can give results for symmetric or symmetrizable first order hyperbolic systems. Obviously, in the case of systems, we have to work more than in the scalar case to obtain the fundamental commutation properties used in J.M. Bony's Theorems proofs.

The matter of this book is organized as follows.

In chapter 1, B. Nadir and A. Piriou give a complete study of the basic conormal spaces $H_s(\Sigma_1 \cup .. \cup \Sigma_r)$. They recall the oscillatory aspect of single conormal progressive waves, and detail some classical examples.

In chapter 2, A. Piriou presents a complete proof of the fundamental J.M. Bony's Theorems of propagation and interaction. Commutation and linearization arguments reduce the proof to results of propagation of Sobolev or Besov regularity for linear systems with non smooth coefficients in the non principal part. These results are obtained thanks to an adaptation of the energy estimate method used in the standard case.

In chapter 3, J.P. Varenne develops a complete symbolic calculus in the case of a single conormal progressing wave for a general system and in the case of transversal interaction of two such waves for 2×2 systems. The general results are applied to classical conormal (without or with logarithms) and piecewise smooth waves.

In chapter 4, H. Bougrini studies the transversal interaction of two single conormal progressing waves for general systems. Using refined conormal spaces $\sum_j I^{\mu_j,\nu_j}(\Sigma_j,\Gamma)$, he improves the conormal regularity in J.M. Bony's Theorems, and obtains the fundamental results on the associated principal symbols.

In chapter 5, G. Laschon considers an example of 4×4 symmetric hyperbolic system with only two characteristic hypersurfaces passing by Γ. He exhibits a solution which is piecewise smooth in the past but presents logarithmic singularities after interaction, revealed thanks to a complete symbolic calculus.

H. Bougrini, G. Laschon, B. Nadir, A. Piriou, J.P. Varenne

Chapter 1

Generalities on conormal distributions.

B. NADIR and A. PIRIOU

1.1 Introduction

To begin with, we recall some general facts about linear partial differential operators. Let Ω be an open set in $\mathbb{R}^n$. The action of $\mathbb{R}^+$ on $\Omega \times (\mathbb{R}^n \setminus 0)$ is defined by $(x, \xi) \longmapsto (x, \lambda\xi)$ for $\lambda > 0$, $x \in \Omega$, $\xi \in \mathbb{R}^n \setminus 0$.
A function $p \in C^\infty(\Omega \times (\mathbb{R}^n \setminus 0)$ is said (positively) homogeneous of degree $\mu \in \mathbb{C}$ if $p(x, \lambda\xi) = \lambda^\mu p(x, \xi)$ for all $\lambda > 0$, $x \in \Omega$, $\xi \in \mathbb{R}^n \setminus 0$. If p is real valued, the hamiltonian field of p is the vector field H_p on $\Omega \times (\mathbb{R}^n \setminus 0)$ defined by $H_p(x, \xi) = \sum_{j=1}^{n} (\frac{\partial p}{\partial \xi_j}(x, \xi) \frac{\partial}{\partial x_j} - \frac{\partial p}{\partial x_j}(x, \xi) \frac{\partial}{\partial \xi_j})$.
For $(y, \eta) \in \Omega \times (\mathbb{R}^n \setminus 0)$ we denote by $t \longrightarrow \chi_t(y, \eta)$ the maximal solution of the ordinary differential system $\frac{d}{dt}(x, \xi) = H_p(x, \xi)$ taking the value (y, η) for $t = 0$. The flow associated to the hamiltonian field H_p of p is defined by χ_t. Remark that p is invariant by χ_t, and that $\chi_t(y, \lambda\eta) = \lambda\chi_s(y, \eta)$ for $\lambda > 0$ and $s = \lambda^{\mu-1}t$ if p is homogeneous of degree $\mu \in \mathbb{R}$. By projecting H_p on the tangent space to the basis Ω, we define

$$\Pi_* H_p(x, \xi) = \sum_{j=1}^{n} \frac{\partial p}{\partial \xi_j}(x, \xi) \frac{\partial}{\partial x_j}$$

Now let L be a $N \times N$ first order linear differential operator in Ω:

$$L = \sum_{j=1}^{n} A_j(x)\, D_{xj} + A(x)$$

where A_j, $A \in C^\infty$ (Ω, End ($\mathbb{C}^N$)), $D_{x_j} = \frac{1}{i}\frac{\partial}{\partial x_j}$.
The principal symbol l of L is the function in $C^\infty(\Omega \times (\mathbb{R}^n \backslash 0)$, End $(\mathbb{C}^N))$ defined by

$$l(x,\xi) = \sum_{j=1}^{n} \xi_j\, A_j(x).$$

We shall note $p(x,\xi) = \det l(x,\xi)$. It is a polynomial in ξ homogeneous of degree N with $C^\infty(\Omega)$ coefficients. We shall always assume that these coefficients are real valued.
The characteristic set $CharL$ of L is the conic subset of $\Omega \times (\mathbb{R}^n \setminus 0)$ defined by $p(x,\xi) = 0$.
If $(y,\eta) \in CharL$, the trajectory of H_p issued from (y,η) is contained in $CharL$, and is called the bicharacteristic of L through (y,η). An open set Ω^- in Ω is said a domain of determinacy of Ω for L if for every (x,ξ) in $CharL$, the bicharacteristic of L issued from (x,ξ) meets $\Omega^- \times (\mathbb{R}^n \setminus 0)$. This obviously imply that $\Pi_* H_p(x,\xi) \neq 0$ for all $(x,\xi) \in CharL$ such that $x \in \Omega \setminus \Omega^-$. The operator L is said of principal type if $\Pi_* H_p(x,\xi) \neq 0$ for all (x,ξ) in $CharL$.
A smooth embedded hypersurface Σ in Ω is a characteristic hypersurface for L if $N(\Sigma) \subset CharL$ where $N(\Sigma)$ is the conormal bundle to Σ deprived of its null section. In this case, the hamiltonian field H_p is tangent to $N(\Sigma)$, and its trajectories are called the bicharacteristics of L on $N(\Sigma)$. Their projections on Σ are called the rays on Σ. Note that for $(x,\xi) \in N(\Sigma)$, $\Pi_* H_p(x,\xi)$ is independant of ξ up to the multiplication by a non vanishing real number. Thus one can define naturally the ray on Σ issued from a point of Σ.
We say that the characteristic hypersurface Σ is simply characteristic for L if $\Pi_* H_p(x,\xi) \neq 0$ where $(x,\xi) \in N(\Sigma)$. In this case the rays on Σ are curves, whose tangent line at a point x is generated by $\Pi_* H_p(x,\xi)$ when $(x,\xi) \in N(\Sigma)$.
Let us now consider a first order $N \times N$ semilinear system:

$$Lu = f(x,u) \quad , \quad f \in C^\infty(\Omega \times \mathbb{C}^N, \mathbb{C}^N). \tag{1.1}$$

Some assumptions are requested in order to give sense to the source term $f(x,u)$ of (1.1), for instance $u \in L^\infty_{loc}(\Omega)$. More precisely, we shall suppose here that u belongs to a Sobolev space $H^{loc}_{s'}(\Omega)$ with $s' > n/2$. Recall that the study in such spaces of diverse problems for (1.1) is well known under general hyperbolicity assumptions for L, specially when L is symmetrizable, and more particulary when L is strictly hyperbolic (see Mizohata [*Mi*] and, for instance, J. Chazarain, A. Piriou [*C.P*], S. Alinhac, P. Gerard [*A.G*]). We shall consider here a given solution u of (1.1) in a space $H^{loc}_{s'}(\Omega)$ with $s' > n/2$, and our goal will be the study of its singularities. When f is linear with respect to u, this study is well known when L is of principal type: the Hörmander's Theorem of propagation of singularities states that for every real number s, the H_s−wave front set $WF_s(u)$ of u is invariant by the flow associated to the hamiltonian field H_p. When f is not linear with respect to u, simple examples show that this invariance is not longer true in general. J.M. Bony proved the invariance when $s < s' + (s' - n/2)$. Such an upper bound of order $2s'$ for the "linear" range of s is generically optimal for semilinear systems of size $N > 2$ (see M. Beals [Be1,2]). When $N = 2$, the convexity properties of $CharL$ allow an upper bound of order $3s'$ (see Beals M. [Be3]). Beyond these bounds, gueninely nonlinear phenomena occur and produce new singularities linked to the geometry of $CharL$, and whose analysis stands on the theory of successive microlocalizations (see Lebeau [L1]). For general results without restriction on the range of s, one must restrain the study to singularities which are sufficiently robust with respect to the nonlinearity.

If E is a vector subspace of $\mathcal{D}'(\Omega)$, the property of propagation of E for (1.1) can be stated as follows:

let Ω^- be a domain of determinacy of Ω for L and E^- the restriction of E to Ω^-. Assume that u is a solution of (1.1) such that

$$u_{|\Omega^-} \in E^-, \quad \text{then} \quad u \in E \tag{1.2}$$

Usually, an iterative method in E equipped with convenient norms allows to reduce this property to the two following:

Invariance of E by nonlinear operations

$$u \in E \Longrightarrow f(x,u) \in E \tag{1.3}$$

Propagation of E for the linear system

$$Lu \in E, \ u_{|\Omega^-} \in E^- \Longrightarrow u \in E \tag{1.4}$$

and to convenient estimates for the corresponding norms.

Consider for instance a closed smooth hypersurface Σ in Ω, a real $s > \frac{n}{2}$

and the space

$$E = \{u \in H_s^{loc}(\Omega) \;/\; WF(u) \subset N(\Sigma)\} \tag{1.5}$$

If $u \in E$ satisfies (1.1), then $Lu = f(x, u)$ belongs also to $H_s^{loc}(\Omega)$. Therefore if $x_0 \in \Sigma$ is such that $(x_0, \xi) \notin CharL$ when $(x_0, \xi) \in N(\Sigma)$, the standard theorem of elliptic regularity shows that $u \in H_{s+1}$ close to x_0, and an iteration of this property gives $u \in C^\infty$ close to x_0. Thus if u is really singular, the hypersurface Σ must be characteristic for L. In this case, the Hörmander's Theorem shows that (1.4) is satisfied, but one can see that (1.5) is not true in general since the inclusion $WF(f(x, u)) \subset N(\Sigma)$ is not warranted when $u \in E$. In the next paragraph we shall consider a subspace of E which satisfies simultaneously (1.4) and(1.5), namely the space $H_s(\Sigma)$ of conormal distributions of H_s^{loc} type with respect to Σ. After having proved in Chapter 2 the propagation of this space, it will be natural to study the interaction of two such conormal solutions correponding to two characteristic hypersurfaces Σ_1, Σ_2 intersecting each other transversally along a smooth manifold Γ of codimension two. We shall need for this the conormal spaces $H_s(\Sigma_1 \cup .. \cup \Sigma_r)$ and $H_s(\Sigma_j, \Gamma)$ associated to the bunch of characteristic hypersufraces $\Sigma_1, .., \Sigma_r$ issued from Γ. All these conormal spaces were introduced by J.M. Bony, and are described in the next paragraph.

In the last paragraph we shall recall the representation of a conormal distribution (in the case of a single hypersurface) by mean of an oscillatory Fourier integral, and the definition of the principal symbol which will have a crucial role in Chapters 3, 4, 5. We shall conclude by the study of some classical examples of conormal distributions.

1.2 Conormal spaces

Let Ω be an open set in $\mathbb{R}^n$ and V a smooth closed submanifold in Ω. We denote by $N(V)$ the conormal bundle to V with the zero section removed. We shall consider several geometrical situations, and denote by Σ one of the following patterns:

$$\Sigma = V \tag{1.6}$$

$$\Sigma = (\Sigma_1, \Gamma) \tag{1.7}$$

where Σ_1 is a smooth closed hypersurface in Ω and Γ a smooth closed hypersurface in Σ_1.

$$\Sigma = \Sigma_1 \cup .. \cup \Sigma_r \tag{1.8}$$

where $r \geq 2$ and $\Sigma_j \quad (1 \leq j \leq r)$ is a smooth closed hypersurface in Ω passing through a smooth submanifold Γ of codimension two in Ω. Furthermore, if $j \neq k$, Σ_j and Σ_k intersect each other transversally and $\Sigma_j \cap \Sigma_k = \Gamma$.
We shall set $N(\Sigma) = N(\Sigma_1) \cup N(\Gamma)$ in the case (1.7), and $N(\Sigma) = N(\Sigma_1) \cup .. \cup N(\Sigma_r) \cup N(\Gamma)$ in the case (1.8). A vector field is said tangent to Σ if it is tangent to Σ_1 and Γ in the case (1.7), tangent to $\Sigma_1, .., \Sigma_r$ (and Γ) in the case (1.8). If $d \in I\!R$, we denote by $\Psi^d(\Omega)$ the space of all propers pseudodifferential operators of order d in Ω and by $\Psi^d_{cl}(\Omega)$ the subspace of all propers classical pseudodifferential operators in Ω. $\mathcal{M}(\Sigma)$ is the Lie algebra of all $M \in \Psi^1_{cl}(\Omega)$, Whose principal symbol vanishes on $N(\Sigma)$.

Definition 1.2.1 *If $s \in I\!R$, we denote by $E_s(\Omega)$ either the Sobolev space $H^{loc}_s(\Omega)$, either the Besov space ${}^\infty H^{loc}_s(\Omega)$. If $k \in I\!N$, we define the conormal space $H_{s,k}(\Sigma)$ as follows:*

$H_{s,0}(\Sigma) = E_s(\Omega)$
$H_{s,k+1}(\Sigma)$ *is the space of all $u \in H_{s,k}(\Sigma)$ such that*

$$Mu \in H_{s,k}(\Sigma) \text{ for every } M \in \mathcal{M}(\Sigma). \tag{1.9}$$

We set $\bigcap_{k \in I\!N} H_{s,k}(\Sigma) = H_{s,\infty}(\Sigma) = H_s(\Sigma)$

Remark that if $u \in H_{s,k}(\Sigma)$ then $WF_{s+k}(u) \subset N(\Sigma)$ where WF_{s+k} is relative to the Sobolev space or the Besov space according to the definition of E_s. In particular $WF(u) \subset N(\Sigma)$ when $u \in H_s(\Sigma)$. It is clear that $\mathcal{M}(\Sigma)$ is a $\Psi^0_{cl}(\Omega)$-module and that $\Psi^d(\Omega)$ operates from $H_{s,k}(\Sigma)$ into $H_{s-d,k}(\Sigma)$

We shall say that a family $(M_j)_{1 \leq j \leq J}$ in $\mathcal{M}(\Sigma)$ generates $\mathcal{M}(\Sigma)$ if:
For every $M \in \mathcal{M}(\Sigma)$, there exists $A_j \in \Psi^0_{cl}(\Omega)$ and $A_0 \in \Psi^0_{cl}(\Omega)$ such that $M = \sum_j A_j M_j + A_0$.

Proposition 1.2.2 *Let ω be an open set relatively compact in Ω. Then, in ω, we have the following properties:*
(i) $\mathcal{M}(\Sigma)$ admits a finite family of generators.
(ii)In the cases (1.6), (1.7) and (1.8) with $r = 2$, $\mathcal{M}(\Sigma)$ is generated by a finite family of vector fields.

Proof. We begin with the proof of (ii). Using a microlocal partition of unity, it is sufficient to check that for every $(x_0, \xi_0) \in \Omega \times (I\!R^n \setminus 0)$ the following property holds:

$$\text{There exists a conic neighbourhood } V \times W \text{ of } (x_0, \xi_0) \tag{1.10}$$

and a finite number of vector fields Z_j *tangent to* Σ *such that: If* $a \in C^\infty(V \times W)$ *is homogeneous of degree 1 and vanishes on* $N(\Sigma)$*, there exists* $a_j \in C^\infty(V \times W)$ *homogeneous of degree 0 with*

$$a = \sum_j a_j z_j, \quad z_j = \text{principal symbol of } Z_j \tag{1.11}$$

If $x_0 \notin \Sigma$, (1.10) is obvious since at least one of the vector fields $Z_j = \partial_{x_j}$ $(1 \leq j \leq n)$ is elliptic at (x_0, ξ_0).
If $x_0 \in \Sigma$ and if moreover $x_0 \notin \Gamma$ (in the cases (1.7) and (1.8) with $r = 2$), we can locally suppose that $\Sigma = \{x = (x_1, x') \in I\!R \times I\!R^{n-1} \;/\; x_1 = 0\}$. We assume here that $codim V = 1$ in the case (1.6), the proof in the general case (1.6) being quite similar. We note $\xi = (\xi_1, \xi')$.
If $\xi_0' \neq 0$, (1.10) is again obvious since at least one of the vector fields $Z_j = \partial_{x_j}$ $(2 \leq j \leq n)$ is elliptic at (x_0, ξ_0).
If $\xi_0' = 0$ using the nullity of $a(x, \xi)$ when $x_1 = \xi' = 0$ and the Taylor's formula, we get $a(x, \xi) = x_1 b_1(x, \xi) + \sum_{j=2}^n \xi_j b_j(x, \xi)$ in a conic neighbourhood of (x_0, ξ_0), where b_1 (resp. b_j for $j \geq 2$) is smooth and homogenous of degree 1 (resp. degree 0). If we set $a_1 = \dfrac{b_1}{i\xi_1}$ and $a_j = \dfrac{b_j}{i}$ when $j \geq 2$, we obtain (1.11) with $Z_1 = x_1 \partial_{x_1}$, $Z_j = \partial_{x_j}$ $(2 \leq j \leq n)$.
Suppose now that $x_0 \in \Gamma$ in the case of (1.7). We can locally suppose that $\Sigma_1 = \{x_1 = 0\}$, $\Gamma = \{x_1 = x_2 = 0\}$. We note $x = (x_1, x_2, x'')$, $\xi = (\xi_1, \xi_2, \xi'')$.
If $\xi_0'' \neq 0$, we obtain easily (1.11) by taking the vector fields $Z_j = \partial_{x_j}$ $(3 \leq j \leq n)$.
If $\xi_0'' = 0$ using the nullity of $a(x, \xi)$ when $x_1 = x_2 = \xi'' = 0$ and Taylor's formula, we can write again

$$a(x, \xi) = x_1 b_1(x, \xi) + x_2 b_2(x, \xi) + \xi'' b''(x, \xi'') \tag{1.12}$$

If $\xi_2^0 \neq 0$, this gives

$$a(x, \xi) = i x_1 \xi_2 a_1(x, \xi) + i x_2 \xi_2 a_2(x, \xi) + \xi'' a''(x, \xi'')$$

with $a_1 = \dfrac{b_1}{i\xi_2}$, $a_2 = \dfrac{b_2}{i\xi_2}$, $a'' = \dfrac{b''}{i}$, thus we obtain (1.11) with $Z_1 = x_1 \partial_{x_2}$, $Z_2 = x_2 \partial_{x_2}$, $Z_j = \partial_{x_j}$ $(3 \leq j \leq n)$.
If $\xi_2^0 = 0$, we have $\xi_1^0 \neq 0$. Thanks to (1.12) and since $a(x, \xi) = b_2(x, \xi) = 0$ if $x_1 = \xi_2 = .. = \xi_n = 0$, we obtain $b_2(x, \xi) = x_1 c_1(x, \xi) + \xi_2 c_2(x, \xi) + \xi'' c''(x, \xi)$, $a = x_1 b_1 + x_2(x_1 c_1 + \xi_2 c_2 + \xi'' c'') + \xi'' b''$, thus (1.11) is satisfied

with $z_1 = x_1\partial_1$, $z_2 = x_2\partial_2$, $z_j = \partial_j$, $(3 \leq j \leq n)$, $a_1 = \dfrac{(b_1 + x_2 c_1)}{i\xi_1}$, $a_2 = \dfrac{c_2}{i}$, $a_j = \dfrac{(b_j + x_2 c_j)}{i}$ for $j \geq 3$.
If $x_0 \in \Gamma$ in the case (1.8) with $r = 2$, the proof of (1.10) is similar: locally we can suppose that $\Sigma_j = \{x_j = 0\}$ for $j = 1, 2$, and the conclusion holds with $z_1 = x_1\partial_1$, $z_2 = x_2\partial_2$, $z_j = \partial_j$, $(3 \leq j \leq n)$. The proof of (ii) is now complete.
For proving (i), we have only to consider the case (1.8) with $r \geq 3$. Let $\sum_{j=1}^{r} \chi_j = 1$ be a microlocal partition of unity with χ_j smooth and homogeneous of degree 0, $\chi_j = 0$ in a conic neighbourhood of $N(\Sigma_i)$ when $i \neq j$. If $M \in \mathcal{M}(\Sigma)$, its principal symbol a vanishes on $N(\Sigma_j) \cup N(\Gamma)$ for a given j in $\{1, .., r\}$. Thus, thanks to the part (ii), we can find a finite number of vector fields $Z_{j,l}$ tangent to Σ_j and Γ, and smooth functions $a_{j,l}$ homogeneous of degree 0, such that $a = \sum_l a_{j,l} z_{j,l}$, where $z_{j,l}$ is the principal symbol of $Z_{j,l}$. Hence $a = \sum_j \chi_j a = \sum_{j,l} a_{j,l} m_{j,l}$, where $m_{j,l} = \chi_j z_{j,l}$ vanishes on $N(\Sigma)$. Thus a finite family of generators of $\mathcal{M}(\Sigma)$ is obtained by taking operators $M_{j,l}$ admitting $m_{j,l}$ as principal symbol.

Remark 1.2.3 *In the conditions of Proposition 1.2.2 (ii), the $C^\infty(\omega)$-module of all vector fields tangent to Σ is also finitely generated.*

The proof is quite similar, and simpler. This property remains true in the case (i), but we will see at the next chapter that the vector fields tangent to $\Sigma_1, .., \Sigma_r$ are not suitable for the study of conormal propagation when $r \geq 3$ (see Remark 2.5.2).

Remark 1.2.4 *If $(M_j)_{1 \leq j \leq J}$ is a family of generators of $\mathcal{M}(\Sigma)$, in the definition 1.2.1 of $H_{s,k}(\Sigma)$ it suffices to ask (1.9) for M belonging to $\{M_1, ..., M_J\}$. Moreover, a distribution u is in $H_{s,k}(\Sigma)$ if and only if $M_1^{\alpha_1} ... M_J^{\alpha_J} u \in E_s(\Omega)$ for all $\alpha_j \in I\!N$ such that $\alpha_1 + .. + \alpha_J \leq k$.*

Proposition 1.2.5 *Let $\Sigma = (\Sigma_1, \Gamma)$ or $\Sigma = \Sigma_1 \cup .. \cup \Sigma_r$ with $r \geq 2$. Let $s \in I\!R$, $k \in I\!N$.*
(i) Let $A \in \Psi^0_{cl}(\Omega)$ such that A is of order $-\infty$ in a conic neighbourhood of $N(\tilde{\Sigma})$ where $\tilde{\Sigma} = \Sigma_1$ if $\Sigma = (\Sigma_1, \Gamma)$, $\tilde{\Sigma} = \Sigma_r$ if $\Sigma = \Sigma_1 \cup .. \cup \Sigma_r$. If $u \in H_{s,k}(\Sigma)$, then $Au \in H_{s,k}(\Sigma')$ where $\Sigma' = \Gamma$ if $\Sigma = (\Sigma_1, \Gamma)$, $\Sigma' = (\Sigma_1 \cup \Gamma)$ if $\Sigma = \Sigma_1 \cup \Sigma_2$, $\Sigma' = \Sigma_1 \cup ... \cup \Sigma_{r-1}$ if $\Sigma = \Sigma_1 \cup .. \cup \Sigma_r$ and $r \geq 3$.
(ii) $H_{s,k}(\Sigma_1 \cup .. \cup \Sigma_r) = \sum_{j=1}^{r} H_{s,k}(\Sigma_j, \Gamma)$ if $r \geq 2$.

Proof. The part (i) is obvious for $k = 0$. Suppose it is true for a $k \in \mathbb{N}$ and consider $u \in H_{s,k+1}(\Sigma)$. Let $M \in \mathcal{M}(\Sigma')$ and $\tilde{A} \in \Psi^0_{cl}(\Omega)$ such that the complete symbol $\sigma_{\tilde{A}}$ of $\tilde{A}$ is equal to 1 in a conic neighbourhood of the conic support of σ_A, and is equal to 0 in a conic neighbourhood of $N(\tilde{\Sigma})$. We can write $MAu = \tilde{A}MAu + Ru$ where $R = (I - \tilde{A})MA$ is regularizing. Since $MA \in \mathcal{M}(\Sigma)$, we have $MAu \in H_{s,k}(\Sigma)$ and the induction hypothesis gives $\tilde{A}MAu \in H_{s,k}(\Sigma')$. Hence $MAu \in H_{s,k}(\Sigma')$ for all $M \in \mathcal{M}(\Sigma')$, which proves that $Au \in H_{s,k+1}(\Sigma')$.

For (ii), let $T \in \Psi^0_{cl}(\Omega)$ such that $\sigma_T = 1$ in a conic neighbourhood of $N(\tilde{\Sigma})$, $\sigma_T = 0$ in a conic neighbourhood of $N(\Sigma_j)$ if $\Sigma_j \neq \tilde{\Sigma}$. If $u \in H_{s,k}(\Sigma)$, then $u = Tu + (I-T)u$, and the part (i) shows that $(I-T)u \in H_{s,k}(\Sigma')$, $Tu \in H_{s,k}(\Sigma_r, \Gamma)$. Hence $H_{s,k}(\Sigma,\Gamma) = H_{s,k}(\Sigma',\Gamma) + H_{s,k}(\Sigma_r,\Gamma)$, and (ii) follows by induction on r.

Before studying nonlinear operations in $H_{s,k}(\Sigma)$, we recall some properties of the Sobolev spaces $H_s(\mathbb{R}^n)$ and the Besov spaces ${}^\infty H_s(\mathbb{R}^n)$, which are denoted by $E_s(\mathbb{R}^n) = E_s$. The corresponding norm is denoted by $\| \cdot \|_s$. See for instance S. Alinhac, P. Gerard [A.G].

Let $u \in E_s$, $v \in E_{s'}$, with $s + s' > 0$. If $t \leq s, s'$ and $t < s + s' - \frac{n}{2}$, then

$$uv \in E_t \text{ and } \| uv \|_t \leq cste \| u \|_s \| v \|_{s'} \tag{1.13}$$

If B is the operator of multiplication by $b \in E_s$, $s > n/2$:

$$\text{B } \textit{is continuous in } E_{s'} \textit{ if } -s < s' \leq s \tag{1.14}$$

Moreover if $A \in OpS^0(\mathbb{R}^n \times \mathbb{R}^n)$, then the commutator $[A, B]$ is continuous from $E_{s'}$ into $E_{s''}$ when:

$$-s < s' \leq s,\ s'' \leq s,\ s'' < s' + Min(1, s - \frac{n}{2}) \tag{1.15}$$

Let f be in $C^\infty(\Omega \times \mathbb{R}^p)$ and $u_1, .., u_p$ be real valued functions in $E_s(\Omega)$ with $s > \frac{n}{2}$, then the composite function

$$g = f(x, u_1, .., u_p) \text{ belongs to } E_s(\Omega), \tag{1.16}$$

and the chain rule

$$\frac{\partial g}{\partial x_j} = \frac{\partial f}{\partial x_j}(x, u_1, .., u_p) + \sum_l \frac{\partial f}{\partial u_l}(x, u_1, .., u_p)\frac{\partial u_l}{\partial x_j} \tag{1.17}$$

is valid. Note that the last products of (1.17) make sense thanks to (1.14) since $s > \frac{n}{2}$, $\frac{\partial f}{\partial u_l}(x, u_1, .., u_p) \in E_s(\Omega)$, $\frac{\partial u_l}{\partial x_j} \in E_{s-1}(\Omega)$.

Now we consider the conormal spaces $H_{s,k}(\Sigma)$.

Proposition 1.2.6 *Let $s > \frac{n}{2}$ and $k \in I\!N$. If $f \in C^\infty(\Omega \times \mathbb{C}^p)$ and $u_1, .., u_p \in H_{s,k}(\Sigma)$, then the composite function $g = f(x, u_1, .., u_p)$ belongs to $H_{s,k}(\Sigma)$*

Proof: We can obviously suppose that the u_j are real valued.
First case: $\Sigma = V$ or $\Sigma = \Sigma_1 \cup \Gamma$ or $\Sigma = \Sigma_1 \cup \Sigma_2$. We proceed by induction on k. The property is known for $k = 0$. If it is true for a given value of k, consider a vector field $Z = \sum_j a_j \partial_{x_j}$ tangent to Σ. If $u \in H_{s,k+1}(\Sigma)$, the chaine rule (1.17) gives

$$Zg = \sum_j a_j \frac{\partial f}{\partial x_j}(x, u_1, .., u_p) + \sum_l \frac{\partial f}{\partial u_l}(x, u_1, .., u_p) Z u_l \tag{1.18}$$

Since $Zu_l \in H_{s,k}(\Sigma)$, the induction hypothesis shows that $Zg \in H_{s,k}(\Sigma)$, hence $g \in H_{s,k+1}(\Sigma)$ thanks to the proposition 1.2.2 (ii).
Second case: $\Sigma = \Sigma_1 \cup ...\Sigma_r$ with $r \geq 3$.
The proof becomes more difficult since we must use pseudodifferential operators $M \in \mathcal{M}(\Sigma)$ and not merely vector fields, thus we cannot longer use the chain rule (1.18). We show at Remark 1.2.13 how this second case can be directly deduced from the first case. But in order to prepare our arguments for chapter 2, we give here a different (and somewhat longer) proof. We begin by giving a substitute of the chain rule. The following definition generalizes the product of two functions.

Definition 1.2.7 *A pseudodifferential product is a finite sum of expressions of the form $(u, v) \longrightarrow C(Au.Bv)$ where $A, B, C \in \Psi^0_{cl}(\Omega)$.*

Suppose now that $(M_j)_{1 \leq j \leq N}$ is a finite family generating $\mathcal{M}(\Sigma)$. Setting $M_0 = I$, we have the following substitute of the Liebniz's formula.

Proposition 1.2.8 *Let π be a pseudodifferential product and $M \in \mathcal{M}(\Sigma)$. Then $M\pi$ is a finite sum of expressions of the form $(u, v) \longrightarrow \pi_\alpha(M_j u, v)+ \pi_\beta(u, M_j v)$, where π_α,π_β are pseudodifferential poducts and $0 \le j \le J$.*

Proof. It suffices to consider the case $\pi(u, v) = uv$. Let

$$I = \sum_1^r T_j, \quad T_j \in \Psi^0_{cl}(\Omega) \tag{1.19}$$

be a partition of the identity such that the complete symbol of T_j vanishes in a conic neighbourhood of $N(\Sigma_k)$ for all $k \neq j$.
Then $M(uv) = \sum_{j,k}^r M(T_j u.T_k v)$. Consider un term $M(T_j u.T_k v)$ for fixed j, k, and remark that $M \in \mathcal{M}(\Sigma) \subset \mathcal{M}(\Sigma_j \cup \Sigma_k)$ if $j \neq k$, $M \in \mathcal{M}(\Sigma) \subset \mathcal{M}(\Sigma_j \cup \Gamma)$ if $j = k$. Thus the Proposition 1.2.2 (ii) shows that M can be expressed modulo $\Psi^0_{cl}(\Omega)$ as a finite sum of terms of the form CZ where $C \in \Psi^0_{cl}(\Omega)$ and Z vector field belonging to $\mathcal{M}(\Sigma_j \cup \Sigma_k)$ or $\mathcal{M}(\Sigma_j \cup \Gamma)$. By Leibniz's formula, we get

$$CZ(T_j u.T_k v) = C(ZT_j u.T_k v) + C(T_j u.ZT_k v)$$

Since ZT_j and $ZT_k \in \mathcal{M}(\Sigma)$ they can be written in the form $\sum_l A_l M_l$, $\sum_l B_l M_l$ with $A_l, B_l \in \Psi^0_{cl}(\Omega)$, and proposition 1.2.8 follows.
Now we give a generalization of the definition 1.2.7.

Definition 1.2.9 *If p is an integer ≥ 1, a p-pseudodifferential product is an expression $\pi_{(p)}$ of the form*

$$\pi_{(p)}(u_1, .., u_p) = \pi(\pi_{(p-1)}(u_1, .., u_{p-1}), u_p)$$

where π is a pseudodifferential product and $\pi_{(1)} = I$.

An induction gives immediately the following generalization of the Proposition 1.2.8.

Proposition 1.2.10 *Let $\pi_{(p)}$ be a p-pseudodifferential product, and $M^\alpha = M_1^{\alpha_1}..M_J^{\alpha_J}$, where $\alpha = (\alpha_1, .., \alpha_J) \in I\!N^J$. Then $M^\alpha \pi_{(p)}(u_1, .., u_p)$ is a finite sum of expressions of the form*

$$\pi_{(p),l}(M^{l_1} u_1, .., M^{l_p} u_p) \tag{1.20}$$

where $\pi_{(p),l}$ is a p-pseudodiffrential product and $l_1, .., l_p \in I\!N^J$, $\mid l_1 \mid +..+ \mid l_p \mid \leq \mid \alpha \mid$, where $\mid \alpha \mid = \alpha_1 + ... + \alpha_J$

Consider now $f \in C^\infty(\Omega \times I\!R^p)$ supported by $\Omega \times K$, K compact in $I\!R^p$. Let $u = (u_1, .., u_p)$ where u_j is real valued if $1 \leq j \leq p$. Denoting the Fourier transform of $f(x,.)$ by $\hat{f}(x,.)$ we get $f(x,u) = (2\pi)^{-p} \int e^{iu(x)\theta} \hat{f}(x,\theta) d\theta$ and

$$M^\alpha f(x,u) = (2\pi)^{-p} \int M^\alpha (e^{iu(x)\theta} \hat{f}(x,\theta)) d\theta \tag{1.21}$$

at least in a weak sense.
Let θ be fixed in $I\!R^p$. Setting $v(x) = u(x).\theta$ and $F(x) = \hat{f}(x,\theta)$, we shall study $M^\alpha(e^{iv}F)$. Using a partition of identity (1.19) with real operators T_j, we obtain $e^{iv}F = \Pi_{j=1}^r e^{iT_j v} F$ and the proposition 1.2.5 gives

$$M^\alpha(e^{iv}F) = \sum_{\text{finite}} \pi_{(r+1),l}(M^{l_1} e^{iT_1 v}, .., M^{l_r} e^{iT_r v}, M^{l_{r+1}} F) \tag{1.22}$$

where $| l_1 | + ... + | l_r | + | l_{r+1} | \leq | \alpha |$.

Consider a term $M^{l_j} e^{iT_j v}$ for a fixed j, $1 \leq j \leq r$. Since $\mathcal{M}(\Sigma) \subset \mathcal{M}(\Sigma_j \cup \Gamma)$, an operator $\tilde{M} \in \mathcal{M}(\Sigma)$ can be expressed modulo $\Psi^0_{cl}(\Omega)$ as a finite sum of terms of the form CZ where $C \in \Psi^0_{cl}(\Omega)$ and Z vector field tangent to Σ_j and Γ. The chain rule gives $CZT(e^{iT_j v}) = iC(e^{iT_j v}.ZT_j v)$. Since $ZT_j \in \mathcal{M}(\Sigma)$, it can be written $\sum_{j=0}^J A_l M_l$ with $A_l \in \Psi^0_{cl}(\Omega)$. This proves that $\tilde{M}(e^{iT_j v}) = \sum_{q=0}^J \Pi_q(e^{iT_j v}, M_q v)$ where Π_q is a pseudodifferential product. Using the Proposition 1.2.10, an easy induction gives

$$M^{l_j} e^{iT_j v} = \sum \Pi_{(\rho+1)}(e^{iT_j v}, M^{m_1} v, .., M^{m_\rho} v) \tag{1.23}$$

where $1 \leq \rho \leq l_j$, $m_j \in I\!N^J$, $| m_1 | + .. + | m_\rho | \leq | l_j |$, and the sum is finite.

Recall that $v(x) = u(x).\theta$, $e^{iT_j v} = e^{i\theta.T_j u}$. Thus (1.23) shows that (1.22) appears as a composite of pseudodifferential products whose arguments are of the form

$$e^{i\theta.T_j u} \ (1 \leq j \leq r), \ \theta^\gamma M^\beta u_{p'} \ (1 \leq p' \leq p), \ M^{l_{r+1}} \hat{f}(x,\theta) \tag{1.24}$$

with $\gamma \in I\!N^p$, $\beta \in I\!N^J$, $l_{r+1} \in I\!N$, $\sum | \beta | + l_{r+1} \leq | \alpha |$.

Let $u \in (H_{s,k}(\Sigma)^p$ and α such $| \alpha | \leq k$. Then all the arguments in (1.24) belong to E_s with polynomial growth in θ (for $e^{i\theta T_j u}$ this results from the lemma 1.2.11 below), the last one being moreover rapidly decreasing in θ. This proves that the integral in (1.21) converges for the

E_s norm, and thus that $f(x,u) \in E_s$. This completes the proof of the proposition 1.2.6. We now state and prove the result we have just used.

Lemma 1.2.11 *Let $w \in E_s$ (which is $H_s(\mathbb{R}^n)$ or ${}^{\infty}H_s(\mathbb{R}^n)$) be real valued and $s > \frac{n}{2}$. Let $\phi \in C_0^{\infty}$. Then the E_s norm of $\phi e^{i\theta w}$ has polynomial growth when $\theta \in \mathbb{R}$*

Proof. We adapt the proof given by Beals [Be] for the invariance of H_s by nonlinear operations when $s > \frac{n}{2}$. Suppose we have found r, μ such that $-s+1 < r \leq s$, and for every $\phi \in C_0^{\infty}$ and $\theta \in \mathbb{R}$:

$$\phi e^{i\theta w} \in E_r \text{ and } \| \phi e^{i\theta w} \|_r \leq C(1+ \mid \theta \mid)^{\mu} \tag{1.25}$$

with a constant C independant of θ.
For $1 \leq j \leq n$ observe that

$$D_j(\phi e^{i\theta w}) = (D_j\phi)e^{i\theta w} + \theta_j \phi e^{i\theta w} D_j w \tag{1.26}$$

and that $D_j w \in E_{s-1}$. Setting $\epsilon = Min(1, \frac{1}{2}(s - \frac{n}{2}))$ and $t = Min(s - 1, r-1+\epsilon)$, we have $t < r$, $t \leq s-1$ and $t < r + (s-1) - \frac{n}{2}$, thus (1.25), (1.13) show that $\phi e^{i\theta w} D_j w \in E_t$ with

$$\| \phi e^{i\theta w} D_j w \|_t \leq C' \| \phi e^{i\theta w} \|_r \| D_j w \|_{s-1} \leq C''(1+ \mid \theta \mid)^{\mu}$$

Applying (1.25) with ϕ replaced by $D_j\phi$ and using (1.26) we deduce that $D_j(\phi e^{i\theta w}) \in E_t$ with a norm in E_t bounded by $C_1(1+ \mid \theta \mid)^{\mu+1}$. This proves that $\phi e^{i\theta w} \in E_{t+1}$ with a norm in E_{t+1} bounded by $C_2(1+ \mid \theta \mid)^{\mu+1}$, that is (1.25) with r replaced by $t+1 = Min(s, r+\epsilon)$ and μ by $\mu+1$. Since (1.25) is satisfied for $r = 0$ in the case of Sobolev spaces, a finite number of such steps proves the Lemma 1.2.11 in the case $E_s = H_s$. From this we deduce that (1.25) is true for $E_r =^{\infty} H_r$ and $r = s - \epsilon'$, $\epsilon' > 0$. The above step gives then (1.25) for $E_s =^{\infty} H_s$.

Remark 1.2.12 *In view of the next chapter it is useful to precise the terms appearing in (1.24):*

If an argument $\theta^{\gamma} M^{\beta} u_{p'}$ is such that $\mid \beta \mid = \alpha$, the others arguments are of the form $e^{i\theta T_j u}$, $\theta^{\gamma'} u_{p''}$, $1 \leq p'' \leq p$, $\hat{f}(x,\theta)$. Therefore one can write $M^{\alpha}(e^{i\theta.u}\hat{f}(x,\theta)) = F_1(x,\theta)) + F_2(x,\theta))$ where F_2 corresponds to the terms of (1.24) such that $\mid \beta \mid \leq \mid \alpha \mid -1$ and F_1 is a finite sum of composite of pseudodifferential products in which one argument is $\theta^\gamma M^\beta u_{p'}$ with $|\beta| = |\alpha|$, $1 \leq p' \leq p$, the others arguments being $e^{i\theta.T_j u}$, $1 \leq j \leq r$, $\theta^{\gamma'} u_{p''}$, $1 \leq p'' \leq p$, $\hat{f}(x,\theta)$.

If $u \in H_{s,k-1}(\Sigma)$ with $k \geq 1$ and $s > \frac{n}{2}$, we have the following linearization of $M^\alpha f(x,u)$ when $|\alpha| = k$:

$$M^\alpha f(x,u) = \sum_{|\beta|=k} T_{\alpha,\beta} M^\beta u + F_\alpha \tag{1.27}$$

where $F_\alpha \in E_s$ and $T_{\alpha,\beta}$ is a linear operator satisfying the two following properties which are easily deduced from (1.14) and (1.15):

$$T_{\alpha,\beta} \text{ is continuous in } E_{s'}(\Omega) \text{ when } -s < s' \leq s \tag{1.28}$$

If $A \in \Psi^0(\Omega)$, the commutator $[A, T_{\alpha,\beta}]$ is continuous from $E_{s'}(\Omega)$ into $E_{s''}(\Omega)$ when

$$-s < s' \leq s, \quad s'' \leq s, \quad s'' < s' + Min(1, s - \frac{n}{2}). \tag{1.29}$$

Remark 1.2.13

In the case $\Sigma = \Sigma_1 \cup .. \cup \Sigma_r$ with $r \geq 3$, if we want only a proof of Proposition 1.2.6 without the precision (1.27), we can use the following simpler arguments:
First step: Algebra property.
Let $u, v \in H_{s,k}(\Sigma)$. Since $u = \sum_{i=1}^r u_i$, $v = \sum_{i=1}^r v_i$ with u_i, $v_i \in H_{s,k}(\Sigma_i, \Gamma)$, we get $uv = \sum_{i=1,j=1}^r u_i v_j$. We know from the first case that $u_i v_j \in H_{s,k}(\Sigma_i, \Gamma) + H_{s,k}(\Sigma_j, \Gamma) \subset H_{s,k}(\Sigma)$ thus $uv \in H_{s,k}(\Sigma)$.
When we restrict to $\omega \subset\subset \Omega$, the space $H_{s,k}$ are equipped with natural norms (up to an equivalence) by taking the E_s norm of the collection $(M^\alpha u), |\alpha| \leq k$ (see the notations in Remark 1.2.4 and Proposition 1.2.10), and the continuity of the multiplication in $H_{s,k}(\Sigma)$ is easy to check.
Second step: Stability under the action of smooth functions.
Let $u = (u_1, .., u_p)$ with $u_j \in H_{s,k}(\Sigma)$. As for (1.21), we have $f(x,u) = (2\pi)^{-p} \int e^{iu(x).\theta} \hat{f}(x,\theta) d\theta$. Thanks to Proposition 1.2.5,(ii), we can write $u = \sum_{i=1}^r v_j$ with $v_j \in H_{s,k}(\Sigma_j, \Gamma)$. Thus

$$f(x,u) = (2\pi)^{-p} \int \Pi_{j=1}^r e^{iv_j(x).\theta} \hat{f}(x,\theta) d\theta. \tag{1.30}$$

We know that $e^{iv_j(x).\theta}$ is in $H_{s,k}(\Sigma_j, \Gamma)$, and it is not hard to establish, thanks to lemma 1.2.11 and the first step, its polynomial growth with respect to θ. Using again the first step, we deduce that the integrand of (1.30) is rapidly decreasing with respect to θ in the space $H_{s,k}(\Sigma)$, thus $f(x,u) \in H_{s,k}(\Sigma)$.

Remark 1.2.14 *In fact, Proposition 1.2.6 of stability of $H_{s,k}(\Sigma)$ by non-linear operations is valid under the following less restrictive hypothesis:*
$s > 1/2,\ s + k > n/2$ *if* $\Sigma = V$
$s > 1,\ s + k > n/2$ *if* $\Sigma = \Sigma_1 \cup \Gamma$ *or* $\Sigma = \Sigma_1 \cup .. \cup \Sigma_r,\ r \geq 2$
(see N. Ritter [R])

1.3 Symbols and examples of conormal distributions

Let X a smooth manifold of dimension n and Y a closed smooth submanifold of codimension l $(1 \leq l \leq n)$ in X. In this paragraph we deal with the conormal space $H_s(Y) = H_{s,\infty}(Y)$ based on the Besov space ${}^{\infty}H_s^{loc}(Y)$. Its definition is quite similar to the one given in Definition 1.2.1, which corresponds to the case $X = \Omega =$ open set in $I\!R^n$. We shall first recall the desription of this space by mean of Fourier integral representations (see L. Hörmander [Hö1,2]). A distribution $u \in \mathcal{D}'(X)$ is in $H_s(Y)$ if and only if u is C^∞ in $X \setminus Y$ and, close to each point of Y, using local coordinates $x = (x_1, .., x_n)$ in X such that $Y = \{x = (x', x") \in I\!R^l \times I\!R^{n-l} \ / \ x' = 0\}$, u can be written in the form

$$u(x) = \int_{I\!R^l} e^{ix'\xi'} a(x", \xi')d\xi', \quad a \in S^\mu(I\!R^{n-l} \times I\!R^l) \tag{1.31}$$

with $\mu = -s - \frac{l}{2}$
Morover, a distribution u of the form

$$u(x) = \int_{I\!R^l} e^{ix'\xi'} b(x, \xi')d\xi', \quad b \in S^\mu(I\!R^n \times I\!R^l) \tag{1.32}$$

can be written in the form (1.30) where

$$a(x", \xi') \sim \sum_{j \geq 0} \left(\frac{< -iD_{x'}, D_{\xi'} >^j}{j!} b(x, \xi')\right)_{|x'=0}$$

It will be useful to point out the degree μ of the symbols a or b appearing in (1.30) or (1.31), and we shall use the notation

$$H_s(Y) = I^\mu(Y), \quad \mu = -s - \frac{1}{2}\text{codim}Y. \tag{1.33}$$

Note that the form (1.30) obviously depends on the choice of the local coordinates ; moreover, even for a given choice, the symbol $a(x'', \xi')$ is not unique in general. To get an invariance property, it is convenient to consider u as a section of the vector bundle $\Omega_{\frac{1}{2}}(X)$ of half densities on X. The local coordinates choosen in (1.30) define local coordinates $(x, \xi) = (x', x''; \xi', \xi'')$ in the cotangent bundle $T^*(X)$ of X. The conormal bundle N(Y) of Y is then parametized by $\mathbb{R}^{n-l} \times \mathbb{R}^l \ni (x'', \xi') \to (0, x''; \xi', 0)$ and the half density on $\mathbb{R}^{n-l} \times \mathbb{R}^l$ defined by

$$(2\pi)^l a(x'', \xi') | \, dx'' \, |^{\frac{1}{2}} | \, d\xi' \, |^{\frac{1}{2}} \tag{1.34}$$

defines a half density valued symbol of degree $\mu + \frac{l}{2}$ on N(Y), which is invariantly defined modulo symbols of degree one lower. This give rise to an isomorphism

$$I^\mu / I^{\mu-1} \to S^{\mu+\frac{l}{2}} / S^{\mu+\frac{l}{2}-1} \tag{1.35}$$

If $u \in I^\mu$, the image by this isomorphism of the equivalence class of u is called the principal symbol of u (note that our normalization factor $(2\pi)^l$ is different from the one in Hörmander [Hö1,2]).

Now we give some classical examples of conormal distributions in the case $X = \mathbb{R}^n$, $Y = \Sigma_1 = \{x_1 = 0\}$. Suppose that u is of the form (1.30) with $x' = x_1$, $x'' = (x_2, ..x_n)$, and $a(x'', \xi_1)$ homogeneous of degree $\alpha \in \mathbb{C}$ for $| \xi_1 |$ large, that is: $a(x'', \lambda\xi_1) = \lambda^\alpha a(x'', \xi_1)$ for $| \xi_1 |$ and $\lambda > 0$ large . We have then $a \in S^\mu$ with $\mu = \Re(\alpha)$, and

$$a(x'', \xi_1) = a^+(x'')\mathcal{H}(\xi_1)\xi_1^\alpha + a^-(x'')\mathcal{H}(-\xi_1)(-\xi)_1^\alpha$$

for $| \xi_1 |$ large, where $a^{\pm} \in C^\infty$, $\mathcal{H}$ = Heaviside function.
Using well known Fourier inverse formulas, we get, modulo C^∞: If $\Re(\alpha) < -1$, $\alpha \notin \mathbb{Z}$, then

$$u(x) \equiv i\Gamma(\alpha+1)(A_+\mathcal{H}(x_1)x_1^{-(\alpha+1)} + A_-\mathcal{H}(-x_1)(-x_1)^{-(\alpha+1)}) \tag{1.36}$$

with $A_{\pm} = a_{\pm}e^{i\pi\frac{\alpha}{2}} - a_{\mp}e^{-i\pi\frac{\alpha}{2}}$.

If $\alpha = -m$ with m integer ≥ 1, then:

$$u(x) \equiv (a_{+} + (-1)^m a_{-})\pi \frac{i^m}{(m-1)!}\mathcal{H}(x_1)x_1^{m-1} -$$

$$(a_{+} - (-1)^m a_{-})(ix_1)^{m-1} ln \mid x_1 \mid \tag{1.37}$$

More generally, one can consider polyhomogeneous symbols $a(x", \xi_1)$, which are asymtotic series of homogeneous symbols with some conditions on the degrees (see L. Hörmander [Hö2]). In particular the functions which are smooth up to the boundary Σ_1 in each (open) side of Σ_1 are called the piecewise smooth functions with respect to Σ_1. These functions are the distributions u of the form (1.30) with $a(x", \xi_1)$ such that

$$a(x", \xi_1) \sim \sum_{j \geq k} a_j(x")/\xi_1^j \quad k \geq 1, a_j \in C^\infty$$

for $\mid \xi_1 \mid$ large.
In this case, note that $u \in I^{-k}(\Sigma_1)$, $u \in C^{k-2}$ if $k \geq 2$, and that $2\pi a_j(x") = i^{-k}[\partial_{x_1}^{j-1}u]_{\Sigma_1}$ if $j \geq k$, where $[.]_{\Sigma_1}$ is the jump across Σ_1 from $x_1 = 0_-$ to $x_1 = 0_+$. In particular, (1.33) gives the principal symbol

$$i^{-k}[\partial_{x_1}^{k-1}u]_{\Sigma_1}\xi_1^{-k} \mid dx" \mid^{\frac{1}{2}} \mid d\xi' \mid^{\frac{1}{2}} \tag{1.38}$$

In chapter 3 we shall consider some special algebras of conormal functions. Here we merely point out that the family of functions of the form (1.35), (1.36) is not stable by multiplication. To get this stability, we must introduce also the function of the form

$$\mathcal{H}(x_1)x_1^d P_{+}(x", lnx_1) + \mathcal{H}(-x_1) \mid x_1 \mid^d P_{-}(x", ln \mid x_1 \mid) \tag{1.39}$$

where $d \in \mathbb{C}$, $\Re(d) \geq 0$, $P_{\pm}(x", .)$ polynomial depending smoothly on $x"$.

We conclude this chapter by checking that these functions are conormal with respect to $\Sigma_1 = \{x_1 = 0\}$.

Proposition 1.3.1 *A function of the form (1.38) belongs to* $I^{-\Re(d)-1+\epsilon}(\{x_1 = 0\})$ *for all* $\epsilon > 0$.

Proof: It is sufficient to consider the case $I\!R^n = I\!R$, and we note $x_1 = x$. It is also sufficient to study the function $u_k(x) = \mathcal{H}(x)x^d(lnx)^k$ with $k \in I\!N$. For $\lambda > 0$, note that
$u_k(\lambda x) = \mathcal{H}(x)\lambda^d x^d \sum_{j+l=k} C_k^j (ln\lambda)^j (lnx)^l$

$$= \lambda^d \sum_{j+l=k} C_k^j (ln\lambda)^j u_l(x).$$

Taking the Fourier transform (it is not hard to prove that the Fourier transform $\hat{u}_k$ of u_k is smooth outside 0) we get

$$\frac{1}{\lambda}\hat{u}_k(\frac{\xi}{\lambda}) = \lambda^d \sum_{j+l=k} C_k^j (ln\lambda)^j \hat{u}_l(\xi), \quad \lambda > 0, \xi \neq 0.$$

Taking successively $\lambda = \xi > 0$ and $-\lambda = \xi < 0$, we deduce that,

$$\hat{u}_k(\xi) = \xi^{-1-d}\hat{u}_k(1) - \textstyle\sum_{0\leq l\leq k-1} C_k^l (ln\xi)^{k-l}\hat{u}_l(\xi) \quad \xi > 0$$

$$\hat{u}_k(\xi) = (-\xi)^{-1-d}\hat{u}_k(-1) - \sum_{0\leq l\leq k-1} C_k^l (ln \mid \xi \mid)^{k-l}\hat{u}_l(\xi), \quad \xi < 0.$$

Then an induction on k gives

$$\hat{u}_k(\xi) = \mathcal{H}(\xi)\xi^{-1-d}P_{k,+}(ln\xi) + \mathcal{H}(-\xi)\mid \xi \mid^{-1-d} P_{k,-}(ln \mid \xi \mid)$$

if $\xi \neq 0$, where $P_{k,\pm}$ are polynomials of degree at most k. The conclusion follows easly.

Chapter 2

Propagation and interaction of conormal waves.

A. PIRIOU

2.1 Introduction

We consider a first order $N \times N$ semilinear system

$$Lu = f(x,u)\ ,\ x \in \Omega \text{ open in } I\!R^n, \tag{2.1}$$

where $L = \sum_{j=1}^{n} A_j(x) D_{x_j} + A(x)$; $A_j,\ A \in C^\infty(\Omega, End(\mathbb{C}^N))$; $D_{x_j} = \frac{1}{i}\frac{\partial}{\partial x_j}$
$f \in C^\infty(\Omega \times \mathbb{C}^N, \mathbb{C}^N)$.
We suppose that the determinant $p(x,\xi)$ of the principal symbol $l(x,\xi) = \sum_{j=1}^{n} \xi_j A_j(x)$ is real when $(x,\xi) \in \Omega \times (I\!R^n \setminus 0)$, and that L is of principal type (see 1.1). The space $H_{s,k}(\Sigma)$ of conormal distributions are based on $E_s(\Omega)$ which is either the Sobolev space $H_s^{loc}(\Omega)$, either the Besov space ${}^\infty H_s^{loc}(\Omega)$, see the Definition 1.2.1.
The following result describes the propagation of a single conormal wave

Theorem 2.1.1 *(J.M. Bony). Let Σ_1 be a smooth closed characteristic hypersurface for L. Let $u \in E_s(\Omega)$, $s > \frac{n}{2}$, satisfying (2.1). Let Ω^- be a domain of determinacy of Ω for L, and $k \in I\!N$.*
If $u_{|\Omega^-} \in H_{s,k}(\Sigma_1 \cap \Omega^-)$, then $u \in H_{s,k}(\Sigma_1)$

We consider now the interaction of two single conormal waves in the following situation: let Σ_1, Σ_2 be two smooth closed characteristic hypersurfaces for L, intersecting each other transversally along a smooth

manifold Γ of codimension two. We assume that
For all $x \in \Gamma$ *and all* $(x, \xi) \in N(\Gamma) \cap Char L$, *the projection*

$$\Pi_* H_p(x, \xi) \textit{ is transversal to } \Gamma \text{ (see 1.1)} \tag{2.2}$$

Then it is easy to see that, close to each point of Γ, there is a finite number of characteristic hypersurfaces for L passing by Γ, and that these hypersurfaces are two by two transversal. We shall assume the following global hypothesis in Ω

$$\Sigma_1, \Sigma_2, .., \Sigma_r \quad (2 \leq r \leq N) \tag{2.3}$$

are smooth closed characteristic hypersurfaces for L, $\Sigma_i \cap \Sigma_j = \Gamma$ *where* $1 \leq i, j \leq r$ *and* $i \neq j$. *Close to each point of* Γ, *the (only) characteristic hypersurfaces through* Γ are $\Sigma_1, \Sigma_2, .., \Sigma_r$.

Then we can state:

Theorem 2.1.2 *(J.M. Bony) Let* Σ_1, Σ_2 *two smooth closed characteristic hypersurfaces for* L, *intersecting each other transversally along a smooth manifold* Γ *of codimension two. We suppose that the hypothesis (2.2), (2.3) are fulfilled, and we note* $\Sigma = \Sigma_1 \cup .. \cup \Sigma_r$ *the bunch of characteristic hypersurfaces issued from* Γ. *Let* Ω^- *be a domain of determinacy of* Ω *for* L, *and* $k \in \mathbb{N}$. *Let* $u \in E_s(\Omega)$, $s > \frac{n}{2}$, *satisfying (2.1). If* Σ_j^- *denote* $\Sigma_j \cap \Omega^-$, $j = 1, 2$ *and*

$$u_{|\Omega^-} \in H_{s,k}(\Sigma_1^- \cup \Sigma_2^-) \textit{ then } u \in H_{s,k}(\Sigma) \tag{2.4}$$

Remark 2.1.3

We could replace the hypothesis (2.4) by $u_{|\Omega^-} \in H_{s,k}(\Sigma \cup \Omega^-)$, but this situation of interaction should be artificial.

Remark 2.1.4

Where $\Gamma \cap \Omega^- = \emptyset$, the hypothesis (2.4) means that $u_{|\Omega^-} \in H_{s,k}(\Sigma_1^-) + H_{s,k}(\Sigma_2^-)$.
When $\Gamma \cap \Omega^- \neq \emptyset$, the hypothesis (2.4) implies that there is no interaction in Ω^-, although it may be geometrically possible. This situation occurs for instance if f(x,.) is linear when x is close to $\Gamma \cap \Omega^-$.

The two Theorems stated before are results of the type (1.2), which stands on (1.3) and (1.4). A crucial step for proving (1.4) is a linear commutation argument (see Proposition 2.2.1 and 2.3.1 below). But the derivation of the corresponding conormal estimates from the classical energy estimates are rather technical (see G. Métivier [Mé] for the case of bounded and L_2^{loc}–conormal solutions). Thus we have preferred to present here the method of J.M. Bony $[Bo1, 2]$, in which the two steps (1.3) and (1.4) are not separated: the commutation is globally performed on the semilinear system (2.1). To deal with the nonlinear source term f(x,u) is very easy when at most two characteristic hypersurfaces are involved. In the general case we simplify the Bony's presentation by using our Proposition 1.2.10 of Leibniz's type. In every case we conclude thanks to an argument of propagation of the E_s regularity for a linear system whose non principal part has some kind of irregularity (see Proposition 2.2.5 and 2.4.1). The paragraph 4 is devoted to the proof of this linear propagation argument, which stands on the usual energy estimates. Our presentation is closely related to the proof of Hörmander's propagation Theorem, and we take the opportunity to give a form of this Theorem for systems (see Theorem 2.4.4)

2.2 Propagation of a single conormal wave

The aim of this paragraph is the proof of Theorem 2.1.1. The conclusion being of local type, it is sufficient to prove it in every $\omega \subset\subset \Omega$ such that $\omega^- = \omega \cap \Omega^-$ is a domain of determinacy of ω for L. Let ω be like this, and let Z_j $(1 \leq j \leq J)$ be generators of the $C^\infty(\omega)$ - module of all vector fields Z in ω tangent to $\Sigma_1 \cap \omega$. (see Remark 1.2.3). We first establish a commutation property.

Proposition 2.2.1 *If Z is a vector field in ω tangent to $\Sigma_1 \cap \omega$, then*

$$[L, Z] = \sum_j C_j Z_j + C_0 + CL \tag{2.5}$$

where C_j, C_0, C are in $C^\infty(\omega, End(\mathbb{C}^N))$ (we identify a scalar operator Z with the differential system Z.Id).

Proof. A partition of unity shows that it suffices to establish (2.5) in a neighbourhood of every point $x_0 \in \omega$. The proof is trivial when $x_0 \notin \Sigma_1$. If $x_0 \in \Sigma_1$, we shall begin by giving a convenient reduction of L.
The following lemma is in fact true for every linear first order differential system L and every simply characteristic hypersurface Σ_1

Lemma 2.2.2 *If $(x_0, \xi_0) \in N(\Sigma_1)$, then the dimension of $kerl(x, \xi)$ is one*

Proof. Let $(x, \xi) \in N(\Sigma_1)$. If $\dim kerl(x_0, \xi_0) \geq 2$, we can choose a basis $e_1, e_2, .., e_N$ of $\mathbb{C}^N$ such that $e_1, e_2 \in kerl(x_0, \xi_0)$. Writing the matrix of $l(x_0, \xi)$ in this basis and computing its determinant $p(x_0, \xi)$ give immediately
$\frac{\partial p}{\partial \xi}(x_0, \xi_0) = 0$, which is impossible.

Lemma 2.2.3 *Close to each point $x_0 \in \Sigma_1$, there exists local coordinates $(x_1, .., x_n)$ in Ω such that:*
$\Sigma_1 = \{x_1 = 0\}$, all the hypersurfaces $\{x_1 = cste\}$ are characteristic for L, and $\Pi_* H_p(x; 1, 0, .., 0) = \frac{\partial}{\partial x_n}$.

Proof. We can suppose $x_0 = 0$, $\Sigma_1 = \{x_1 = 0\}$. Let $\xi_0 = (1, 0, .., 0)$. We know that $\Pi_* H_p(0, \xi_0)$ is tangent to Σ_1 and different from 0. Thus we can suppose that $\frac{\partial p}{\partial \xi_n}(0, \xi_0) \neq 0$. For $y \in \mathbb{R}^{n-1}$ close to 0, let $\xi_n = \xi_n(y)$ the local solution of $p(y, 0; 1, 0, .., 0, \xi_n) = 0$ such that $\xi_n(0) = 0$. For $t \in \mathbb{R}$ close to 0, let $x = x(y, t)$, $\xi = \xi(y, t)$ be the solution of the hamiltonian system:

$$\frac{\partial x}{\partial t} = \frac{\partial p}{\partial \xi}(x, \xi), \quad \frac{\partial \xi}{\partial t} = -\frac{\partial p}{\partial x}(x, \xi) \tag{2.6}$$

such that $x(y, 0) = (y, 0)$, $\xi(y, 0) = (1, 0, ..., 0, \xi_n(y))$. Let φ be the local solution of the non characteristic Cauchy problem

$$p(x, \varphi'(x)) = 0\ , \quad \varphi = x_1 \text{ when } x_n = 0, \ \varphi'(0) = \xi_0$$

We know that φ is constant along the solutions of (2.6) with fixed y, therefore $\varphi(x) = y_1$ when $x = x(y, t)$. The new coordinates $(y_1, .., y_{n-1}, t)$ satisfy the conditions stated in the Lemma.

Lemma 2.2.4 *Let the local coordinates be choosen like in the Lemma 2.2.3. Then, taking a suitable basis of $\mathbb{C}^N$ (depending smoothly on x in a neighbourhood of x_0) and after left multiplication of L by a suitable element of $GL\ \mathbb{C}^N$ depending smoothly on x, one can write*

$$L = \sum_{j=1}^{n} A_j(x)\, \partial_{xj} + A(x), \tag{2.7}$$

where

$$A_1(x) = \begin{pmatrix} 0 & 0 \\ 0 & I_{N-1} \end{pmatrix},$$

$$I_{N-1} = \textit{Identity matrix of dimensions } N-1$$

Proof. Thanks to the lemma 2.2.2 and since all the hypersurfaces $\{x_1 = cste\}$ are characteristic for L, L is of the form (2.7) with $dimker A_1(x) = 1$. We choose a basis $e_1(x)$ of this kernel depending smoothy on x in a neighbourhood of x_0, and we complete it into a basis of $\mathbb{C}^N$. Denoting by $V_1(x), .., V_N(x)$ the columns of the matrix of $A_1(x)$ in this basis, we have $V_1(x) = 0$ and $V_2, .., V_N$ are independant. Thus, possibly after a permutation of two rows (that is left multiplication by a suitable invertible matrix) we obtain a matrix of the form $\begin{pmatrix} 0 & B(x) \\ 0 & M(x) \end{pmatrix}$ where M(x) is a square invertible matrix of dimensions $N-1$.

Left multiplication by $\begin{pmatrix} 1 & -BM^{-1} \\ 0 & M^{-1} \end{pmatrix}$ gives the wanted form for $A_1(x)$

End of the proof of proposition 2.2.1. We can assume that the properties stated in the Lemma 2.2.4 are fulfilled, and that $Z_1 = x_1\partial x_1$, $Z_j = \partial_{x_j}$ $(2 \leq j \leq n)$. The term $A(x)$ of (2.7) can be obviously neglected. For $2 \leq j \leq n)$, we have $[iL, Z_j] = -\frac{\partial A_1}{\partial x_1}\partial_{x_1} - \sum_{k=2}^{n} \frac{\partial A_k}{\partial x_j}\partial x_k = -\sum_{k=2}^{n} \frac{\partial A_k}{\partial x_j} Z_k$ since the matrix A_1 is constant. Moreover, we have

$$[iL, Z_1] = A_1\partial_{x_1} - \sum_{k=2}^{n} x_1 \frac{\partial A_k}{\partial x_1} Z_k - \frac{\partial A_k}{\partial x_1} Z_1.$$

Since $A_1\partial x_1 = L - \sum_{k=2}^{n} A_k \partial_{x_k}$,we get

$$[iL, Z_1] = -\sum_{k=2}^{n}(A_k + x_1\frac{\partial A_k}{\partial x_1})Z_k + iL$$

End of the proof of Theorem 2.1.1. Suppose that for a given l such that $1 \leq l \leq k$ we have

$$u \in H_{s,l-1}(\Sigma \cap \omega) \tag{2.8}$$

Using the notations given at the begining of this paragraph, let $U^{(l)}$ be the collection of $Z^\alpha u = Z_1^{\alpha_1}..Z_j^{\alpha_j} u$ when $|\alpha| = \alpha_1 + .. + \alpha_j \leq l$. Starting from Proposition 2.2.1, an easy induction gives

$$[L, Z^\alpha] = \sum_{|\beta| \leq |\alpha|} C_{\alpha,\beta} Z^\beta + \sum_{|\gamma| \leq |\alpha|-1} \widetilde{C}_{\alpha,\gamma} Z^\gamma L \tag{2.9}$$

where $C_{\alpha,\beta}$ and $\widetilde{C}_{\alpha,\gamma} \in C^\infty(\omega, \text{ End }(\mathbb{C}^N))$. From (1.18) we get also

$$Z^\alpha f(x,u) = \sum_{j=1}^{N} \frac{\partial f}{\partial_{u_j}}(x,u) Z^\alpha u + g_\alpha(x, U^{l-1}) \text{ if } |\alpha| = l \tag{2.10}$$

where g_α is a smooth function.
From (2.1) we have $Z^\alpha Lu = Z^\alpha f(x,u)$, thus $LZ^\alpha u = Z^\alpha f(x,u) + [L, Z^\alpha]u$.
Using (2.10), (2.9), and remarking that $Z^\gamma Lu = Z^\gamma f(x,u) \in E_s(\omega)$ where $|\gamma| \leq l-1$ and that $g_\alpha(x, U^{l-1}) \in E_s(\omega)$, we see that $U^{(l)}$ satisfies a system of the form

$$LU_i^{(l)} + \sum_j a_{i,j}\, U_j^{(l)} = f_i \tag{2.11}$$

where the coefficient $a_{i,j}$, f_i belong to $E_s((\omega))$.
We know that $U^{(l)}_{|\omega^-} \in E_s(\omega^-)$, thus (2.11) implies that $U^{(l)} \in E_s((\omega))$ thanks to the Proposition 2.2.5 stated below. Thus we have established that (2.8) implies that $U \in H_{s,l}(\omega)$.
Since (2.8) is obviously fulfilled if $l = 1$, the proof of Theorem 2.2.1 is complete.
We now state the linear propagation result we have just used, and which will be proved in the paragraph 4(see Proposition 2.4.1).

Proposition 2.2.5 *Let L be $N \times N$ first order differential system in $\omega \subset \mathbb{R}^n$. we assume that the determinant p of the principal symbol of L is real, and that L is of principal type. Let $s > n/2$ and $U_i (1 \leq i \leq I)$ be $\mathbb{C}^N$ valued functions belonging to $E_{s-1}(\omega)$ and satisfying a linear differentiel system of the form.*

$$LU_i + \sum_j a_{ij} U_j = f_i \ \ (1 \leq i \leq I) \tag{2.12}$$

where the coefficients a_{ij} and f_i belong to $E_s(\omega)$. Let ω^- be a domain of determinacy of ω for L.
If $U_{i_{|\omega^-}} \in E_s(\omega^-)$ when $1 \leq i \leq I$, then $U_i \in E_s(\omega)$ when $1 \leq i \leq I$.

We finish this paragraph by giving a property related to lemma 2.2.2 and to the so called polarization property associated to Σ_1 (see P.D. Lax[La]), which will be useful later

Proposition 2.2.6 *Let L be a $N\times N$ first order linear differential system ; denote by l its principal symbol and assume that $p = detl$ is real. Let Σ_1 be a smooth simply characteristic hypersyrface for L. If $e_1(x)$, $e_1^{'}(x)$ are smooth non null vectors in $\mathbb{C}^N$ such that $e_1(x) \in kerl(x,\xi)$ and $e_1^{'}(x) \in ker^t l(x,\xi)$ when $(x,\xi) \in N(\Sigma_1)$, we define the first order scalar linear differential operator L_1 by $L_1(v) =< e_1^{'}, L(ve_1) >$. Let $l_1 =< e_1^{'}, le_1 >$ be the principal symbol of L_1. Then, on $N(\Sigma_1)$, the hamiltonian field of l_1 is proportional to the hamiltonian field H_p of p.*

Proof. Notice that for another choice of vectors $e_1(x), e_1^{'}(x)$, Lemma 2.2.2 shows that, on $N(\Sigma_1)$, L_1 is multiplied by a non vanishing smooth function. Moreover, if $T(x)$ is a smooth $N\times N$ invertible matrix and if L is replaced by $T(x)L$, we can replace $e_1^{'}(x)$ by ${}^tT^{-1}(x)l_1^{'}(x)$ and leave $e_1(x)$ unchanged, which leave the operator L_1 unchanged. Therefor, we can (locally) assume that the properties stated in Lemma 2.2.3 and Lemma 2.2.4 are satisfied. We use the notations $\xi = (\xi^{'}, \xi_n)$, $\xi_0^{'} = (1,0,..,0) \in \mathbb{R}^{n-1}$, $\xi_0 = (\xi_0^{'}, 0) = (1,0,..,0) \in \mathbb{R}^n$, $y = (x,\xi^{'})$, $y_0 = (x_0, \xi_0^{'})$.
Since $p(x_0,\xi_0) = 0$ and $\dfrac{\partial p}{\partial \xi_n}(x_0,\xi_0) = 1$, the equation $p(x;\xi^{'},\xi_n) = 0$ admits locally a smooth solution $\xi_n = \lambda(y)$ such that $\lambda(y_0) = 0$, and we have $p(x,\xi) = (\xi_n - \lambda(y))q(x,\xi)$ where $q(x_0,\xi_0) = 1$. Let $e(y)$, $e^{'}(y)$ be respectively basis of $kerl(y,\lambda(y))$ and $ker^t l(y,\lambda(y))$ such that $e(y_0) = e^{'}(y_0) = (1,0,..,0)$. Since $l(y,\lambda(y))e(y) = 0$, we get
$$(\frac{\partial l}{\partial y_j}(x_0,\xi_0) + \frac{\partial l}{\partial \xi_n}(x_0,\xi_0)\frac{\partial \lambda}{\partial y_j}(y_0))e(y_0) + l(x_0,\xi_0)\frac{\partial e}{\partial y_j}(y_0) = 0.$$
But $\dfrac{\partial \lambda}{\partial y_j}(y_0) = -\dfrac{\partial p}{\partial y_j}(x_0,\xi_0)$,
$$< e^{'}(y_0), l(x_0,\xi_0)\frac{\partial e}{\partial y_j}(y_0) >=<^t l(x_0,\xi_0)e^{'}(y_0), \frac{\partial e}{\partial y_j}(y_0) >= 0,$$
and similarly
$$\frac{\partial l_1}{\partial y_j}(x_0,\xi_0) =< e^{'}(y_0), \frac{\partial l}{\partial y_j}(x_0,\xi_0)e(y_0) >,$$
thus $\dfrac{\partial l_1}{\partial y_j}(x_0,\xi_0) = \mu\dfrac{\partial p}{\partial y_j}(x_0,\xi_0)$,
where $\mu =< e^{'}(y_0), \dfrac{\partial l}{\partial \xi_n}(x_0,\xi_0)e(y_0) >$.
Because of the obvious equality
$\dfrac{\partial l_1}{\partial \xi_n}(x_0,\xi_0) = \mu\dfrac{\partial p}{\partial \xi_n}(x_0,\xi_0)$, we have just to check that $\mu \neq 0$, which follows easily from the form of A_1 given in the lemma 2.2.4: since $e(y_0) = e^{'}(y_0) = (1,0,..,0)$, μ is the top left element of $A_n(x)$, but $p(x_0;\xi_0^{'},\xi_n) =$

$det(A_1(x_0)+\xi_n A_n(x_0))$, hence $\mu = \frac{\partial p}{\partial \xi_n}(x_0;\xi_0',0) = 1$.

2.3 Interaction of two transversal conormal waves

This paragraph is devoted to the proof of Theorem 2.1.2. As in the previous paragraph, we consider $\omega \Subset \Omega$ such that $\omega^- = \omega \cap \Omega^-$ is a domain of determinacy of ω for L ; Let M_j $(1 \le j \le J)$ be generators of $\mathcal{M}(\Sigma \cup \omega)$ - see Proposition 1.2.2.
The commutation property is now the following

Proposition 2.3.1 *If $M \in \mathcal{M}(\Sigma \cup \omega)$, then*

$$[L, M] \equiv \sum_j C_j M_j + C_0 + CL \tag{2.13}$$

where C_j, C_0, C are pseudodifferential systems belonging to $\Psi^0_{cl}(\omega)$. The notation $A \equiv 0$ means that the complete symbol of A vanishes in a conic neighbourhood of $N(\Gamma \cap \omega)$.

A partition of unity shows again that it is sufficient to establish (2.13) in a neighbourhood of every point $x_0 \in \omega$. The case $x_0 \notin \Sigma$ is trivial, and the case $x_0 \in \Sigma \setminus \Gamma$ results from Proposition 2.2.1. It remains to prove (2.13) close to a point $x_0 \in \Gamma$. For this purpose, we shall give a reduction of L by considering a basis of $\mathbb{C}^N$ linked to the polarization defined by Σ_j, $1 \le j \le r$ (see Proposition 2.2.6).

Reduction 2.3.2

We choose local coordinates such that $x_0 = 0$, $\Gamma = \{x_1 = x_2 = 0\}$, $\{x_n = 0\}$ non characteristic for L. Left multiplication of L by suitable invertible matrix allows to write $L = \sum_{j=1}^n A_j(x) D_{x_j} + A_0(x)$ with $A_n(x) = I$. We use the notation $x = (x_1, x'', x_n)$ and we identify $(0, x'', 0)$ with x''.
Hypothesis (2.2), (2.3) show that the equation $det(A_1(x'') + \lambda I) = 0$ admits r simple and real roots $\lambda = \lambda_j(x'')$, $1 \le j \le r$, the fiber of $N(\Sigma_j)$ at x'' being generated by $(1, 0, .., 0, \lambda_j(x''))$. We consider a basis $e_j(x) \in \mathbb{C}^N$ of Ker$l(x,\xi)$ when $(x,\xi) \in N(\Sigma_j)$, depending smoothly on $x \in \Sigma_j$. Then $e_j(x'')$ generates the eigenspace of $A_1(x'')$ corresponding to the eigenvalue $-\lambda_j(x'')$. Denote by $(e_{r+1}(x''), .., e_N(x''))$ a basis (depending smoothly of x'') of the sum of characteristic spaces of $A_1(x'')$ corresponding to the non real eigenvalues of $A_1(x'')$. Then $(e_1(x''), .., e_N(x''))$ is a basis of $\mathbb{C}^N$. We

extend $e_j(x)$ $(1 \leq j \leq r)$ outside Σ_j and $e_j(x'')$ $(r+1 \leq j \leq N)$ outside Γ in such a way that $(e_1(x), .., e_N(x))$ remains a basis of $\mathbb{C}^N$ depending smoothly on x.
We construct in the same way a basis $(e_1'(x), .., e_N'(x))$ of $\mathbb{C}^N$ assciated to the transposed ${}^tl$ of l.
If we set $u = \sum_{j=1}^N u_j e_j$, the system (2.1) is equivalent to

$$\sum_{j=1}^{N} L_{j,k} u_k = f_j(x, u) \quad (1 \leq j \leq N) \tag{2.14}$$

where $L_{j,k}$ is the first order scalar differential defined by

$$L_{j,k} v = < e_j', L(v e_k) > \quad \text{and } f_j = < e_j', f > . \tag{2.15}$$

The principal symbol $l_{j,k}$ of $L_{j,k}$ is

$$l_{j,k}(x, \xi) = < e_j'(x), l(x, \xi) e_k(x) > = <^t l(x, \xi) e_j'(x), e_k(x) > \tag{2.16}$$

For proving (2.13) in the neighbourhood of $x_0 \in \Gamma$, we can replace L by the matrix $(L_{j,k})$. We give now some properties of this matrix

Proposition 2.3.3 *(i) For $1 \leq j \leq r$ and $1 \leq k \leq N$, $l_{j,k}$ and $l_{k,j}$ are null on $N(\Sigma_j)$.*
(ii) If $1 \leq j, k \leq N$, $j \neq k$, j or $k \leq r$, then $l_{j,k}$ is null on $N(\Gamma)$.
(iii) Let $L' = (L_{j,k})_{r+1 \leq j,k \leq N}$. Then L' is elliptic on $N(\Gamma)$.
(iv) If $1 \leq i \leq r$, let $L_i = (L_{j,k})_{1 \leq j,k \leq N; j \neq i; k \neq i}$. Then L_i is elliptic on $N(\Sigma_i)$.
(v)If $1 \leq j \leq r$, then the hamiltonian field of $l_{j,j}$ is proportional to the hamiltonian field H_p of p on $N(\Sigma_j)$.

Proof. (v) results from Proposition 2.2.6 since L is of principal type.
(i) results from (2.16).
If $1 \leq j, k \leq r$ and $j \neq k$, (i) shows that $l_{j,k}$ vanishes on $N(\Sigma_j)$ and $N(\Sigma_k)$, thus $l_{j,k}$ vanishes also on $N(\Sigma_j)_{|\Gamma} + N(\Sigma_k)_{|\Gamma} = N(\Gamma)$ since $l_{j,k}(x, .)$ is linear. Suppose now that $1 \leq j, k \leq N$, $j \leq r$, $k > r$. Let $x = (0, x'', 0) \in \Gamma$ and $\xi = (1, 0, .., 0, \lambda)$ when $\lambda \in \mathbb{R}$. We set

$$F(\lambda) = l_{j,k}(x, \xi) = < e_j'(x), l(x, \xi) e_k(x) >, \; l(x, \xi) = A_1 x + \lambda I \tag{2.17}$$

Recall that $e_k(x)$ belongs to V, the sum of the characteristic spaces of $A_1(x)$ corresponding to the non real eigenvalues of $A_1(x)$. We have also

$l(x,\xi)e_k(x) \in V$. Since $e_j^{'}(x)$ is an eigenvector of ${}^tA_1(x)$ corresponding to the real eigenvalue $-\lambda_j(x)$ of $A_1(x)$, we get

$$< e_j^{'}(x), l(x,\xi)e_k(x) >= 0,$$

thus the function F defined in (2.17) is null, which concludes the proof of (ii). Moreover, for the same values of x and ξ as before, we have $p(x,\xi) = \tilde{p}(x,\xi)\Pi_{j=1}^{r} l_{j,j}(x,\xi)$ where $\tilde{p}$ is the determinant of the principal symbol of $L^{'}$. But $l_{j,j}(x,\xi) = 0$ when $\lambda = \lambda_j(x)$, thus $\tilde{p}(x,\xi) \neq 0$ for all $\lambda \in I\!R$ since the real zeros λ of $p(x,\xi)$ are $\lambda_1(x), \,..\, , \lambda_r(x)$ and these zeros are simple. Hence (iii) is proved.
For (iv), we develop $p = det l$ with respect to the i-th column:
$p = \sum_{s=1}^{N} l_{si}C_{si}$, where C_{si} is the cofactor (with sign) of the term l_{si} in matrix $(l_{j,k})_{1\leq j,k\leq N}$. On $N(\Sigma_i)$, (i) proves that $l_{si} = 0$ when $1 \leq s \leq N$; we have also $C_{si} = 0$ when $s \neq i$ as seen when developing the determinant of $(l_{j,k})_{j\neq s, k\neq i}$ with respect to the row of index $j = i$. Therefore we get

$$\frac{\partial p}{\partial \xi_k} = (\frac{\partial l_{ii}}{\partial \xi_k})C_{ii} \text{ on } N(\Sigma_i), \;\; 1 \leq k \leq N. \tag{2.18}$$

which implies that $C_{ii} \neq 0$ on $N(\Sigma_i)$, that is (iv). Remark that we have obtained another proof of (v).

End of the proof of Proposition 2.3.1.
With the notations of Proposition 2.3.3, let $F \in \Psi_{cl}^{-1}$ be a microlocal parametrix of $L^{'}$ in a conic neighbourhood of $N(\Gamma)$. Let $E \in \Psi_{cl}^{-1}$ be an elliptic diagonal $r \times r$ system. Then

$$\begin{pmatrix} E & 0 \\ 0 & F \end{pmatrix} L = \tilde{L} = (\tilde{L}_{j,k})_{1\leq j,k\leq N}$$

belongs to Ψ_{cl}^{0} and satisfies the same properties as the properties stated for L at Proposition 2.3.3, and moreover (iii) is refined into $\tilde{L}^{'} = I + R$, $R \equiv 0$. For proving (2.13), we have only to establish an equality of the form

$$[\tilde{L}, M] \equiv A + B\tilde{L} \tag{2.19}$$

where A , B are $N \times N$ systems in Ψ_{cl}^{0}, and $A \in \mathcal{M}^0(\Sigma)$, which means that the principal symbol of A vanishes on $N(\Sigma)$.
Since $[\tilde{L}^{'}, M] = [R, M] \equiv 0$, we have only to consider $[\tilde{L}_{j,k}, M]$ when j or $k \leq r$. Let $I = \sum_{l=1}^{r} \chi_l + \chi_0$ be a pseudodifferential partition of identity

such that: when $1 \leq l \leq r$, $\chi_l \in \Psi^0_{cl}$, the principal symbol of χ_l is one in a conical neighbourhood of $N(\Sigma_l)$ and zero in a conical neighbouhood of $N(\Sigma_{l'})$ if $l \neq l'$; $\chi_0 \in \Psi^0_{cl}$, the principal symbol of χ_0 is zero in a conical neighbourhood of $N(\Sigma_j)$ if $1 \leq j \leq r$. We have then

$$[\tilde{L}_{j,k}, M] = \sum_{l=0}^{r} [\tilde{L}_{j,k}, M]\chi_l$$

(1)Study of $[\tilde{L}_{j,k}, M]\chi_l$ when $1 \leq j, k \leq N$, j or $k \leq r$, $j \neq k$.

Case $l = 0$. Since the principal symbols of $\tilde{L}_{j,k}$ (see Proposition 2.3.3 (ii)) and M vanish on the lagrangian manifold $N(\Gamma)$, the same is true for the commutator $[\tilde{L}_{j,k}, M]$. Therefore $[\tilde{L}_{j,k}, M]\chi_0 \in \mathcal{M}^0(\Sigma)$.
Case $1 \leq l \leq r$. As just seen, the principal symbol of $[\tilde{L}_{j,k}, M]$ vanishes on $N(\Sigma)$. It vanishes also on $N(\Sigma_l)$ if $l = j$ or k. Since the principal symbol of χ_l vanishes on $N(\Sigma_{l'})$ when $l \neq l'$, we obtain that $[\tilde{L}_{j,k}, M]\chi_l \in \mathcal{M}^0(\Sigma)$ when j or $k = l$. Suppose now that j and k are different from l. If $u = (u_1, .., u_N)$ is a function, we use the notations $\tilde{L}u = f$, $u^l = (u_i)_{1 \leq i \leq N, i \neq l}$. Let E_l be a microlocal parametrix of L_l on $N(\Sigma_l)$ (see Proposition 2.3.3, iv). We have

$$L_l u^l = f^l - h^l, \quad \text{where } h_i = \tilde{L}_{i,l} u_l \ \ (1 \leq i \leq N).$$

Thus

$$u^l = E_l f^l - E_l h^l + R_l u^l, \quad R_l \in Reg(\Sigma_l),$$

which means that the complete symbol of R_l vanishes in a conic neighbourhood of $N(\Sigma_l)$. Taking the componant of index k in u^l (recall that $k \neq l$), we get

$$u_k = \sum_{i \neq l} E_{k,i} f_i - \sum_{i \neq l} E_{k,i} \tilde{L}_{i,l} u_l + \sum_{i \neq l} R_{k,i} u_i$$

with $E_{k,i}$, $R_{k,i} \in \Psi^0_{cl}$ and $R_{k,i} \in Reg(\Sigma_l)$. Denoting $[\tilde{L}_{j,k}, M]\chi_l = A$, we deduce that

$$Au_k = \sum_{i \neq l} AE_{k,i} f_i - \sum_{i \neq l} AE_{k,i} \tilde{L}_{i,l} u_l + \sum_{i \neq l} AR_{k,i} u_i$$

The principal symbols of $[\tilde{L}_{j,k}, M]$, χ_l, $\tilde{L}_{i,l}$ vanish respectively on $N(\Gamma)$ (since $j \neq k$), $N(\Sigma_{l'})$ for $l' \neq l$, and $N(\Sigma_l)$. Thus the product $AE_{k,i}\tilde{L}_{i,l}$ belongs to $\mathcal{M}^0(\Sigma)$, and the same property is true for $AR_{k,i}$. Summing up, and remarking that nothing is changed if we replace $[\tilde{L}_{j,k}, M]$ by $\tilde{L}_{j,k}$, we have proved the following lemma.

Lemma 2.3.4 *Let j, k such that $1 \leq j, k \leq N$, $j \neq k$, j or $k \leq r$. Then there exists $1 \times N$ systems $A_{j,k}$, $B_{j,k}$ of operators in Ψ^0_{cl} such that $A_{j,k} \in \mathcal{M}^0(\Sigma)$ and, for all $u = (u_1, .., u_N)$:*

$$[\tilde{L}_{j,k}, M]u_k = A_{j,k}u + B_{j,k}\tilde{L}u. \tag{2.20}$$

The same type of equality is true if $[\tilde{L}_{j,k}, M]$ is replaced by $\tilde{L}_{j,k}$

(2)Study of $[\tilde{L}_{j,j}, M]\chi_l$ when $1 \leq j \leq r$.

The only difference with the first case is that the principal symbol of $\tilde{L}_{j,j}$ is not zero on $N(\Gamma)$.
But we have the following property.

Lemma 2.3.5 *If $1 \leq j \leq r$, there exists $C_j \in \Psi^0_{cl}$ such that the principal symbol of $[\tilde{L}_{j,j}, M] - C_j\tilde{L}_{j,j}$ is zero on $N(\Gamma)$.*

We have then $[\tilde{L}_{j,j}, M]u_j = ([\tilde{L}_{j,j}, M] - C_j\tilde{L}_{j,j})u_j + C_j\tilde{L}_{j,j})u_j$ and Lemma 2.3.4 remains true for $j = k$ if we replace in (2.20) $[\tilde{L}_{j,j}, M]$ by $[\tilde{L}_{j,j}, M] - C_j\tilde{L}_{j,j}$. On the other hand, if $\tilde{L}u = f$, we have $\tilde{L}_{j,j}u_j = f_j - \sum_{j\neq k}\tilde{L}_{j,k}u_k$ and we can apply Lemma 2.3.4 to each term $\tilde{L}_{j,k}u_k$, $j \neq k$. Finally we get an equality of the type (2.20) when $j = k \in \{1, .., r\}$.

The steps (1) and (2) given above prove (2.19) and Proposition 2.3.1. It remains to establish Lemma 2.3.5.
Proof of the Lemma 2.3.5. We choose local coordinates such that $\Sigma_j = \{x_1 = 0\}$, $\Gamma = \{x_1 = x_n = 0\}$, and we note $x = (x_1, x'', x_n)$. Recall that the principal symbol of $\tilde{L}_{j,j}$ is of the form $el_{j,j}$ where e is an elliptic homogeneous symbol of degree -1 and $l_{j,j}$ is the principal symbol of $L_{j,j}$. Thanks to Proposition 2.3.3(v) and to Hypothesis (2.2) of transversality, we have

$$l_{j,j}(0, x'', 0; 0, 0, \xi_n) = a(x'')\xi_n \text{ with } a(x'') \neq 0.$$

On the other hand, the principal symbol σ of $[\tilde{L}_{j,j}, M]$ vanishes on $N(\Sigma_j)$, thus $\sigma(0, x'', 0; \xi_1, 0, \xi_n) = 0$ when $\xi_n = 0$. It follows that $\rho(x'', \xi_1, \xi_n) = \sigma(0, x'', 0; \xi_1, 0, \xi_n)/\xi_n$ is a smooth homogeneous symbol of degree -1. We extend it into a smooth homogeneous symbol $\tilde{\rho}(x, \xi)$ of degree -1, and take $C_j \in \Psi^0_{cl}$ with principal symbol $\tilde{\rho}(x, \xi)/a(x'')e(x, \xi)$.

We are now ready to prove Theorem 2.1.2
Using the notations given at the beginning of this paragraph, let U^l be

the collection of $M^\alpha u = M_1^{\alpha_1}..M_J^{\alpha_J} u$ when $|\alpha| = \alpha_1 + ..\alpha_J \leq l$. Suppose that, for a l such that $1 \leq l \leq k$, we have

$$u \in H_{s,l-1}(\Sigma \cap \omega) \tag{2.21}$$

Using the same method as in the proof of Theorem 2.1.1, we shall deduce that $u \in H_{s,l}(\Sigma \cap \omega)$. Instead of (2.9), we have now from Proposition 2.3.1

$$[L, M^\alpha] \equiv \sum_{|\beta| \leq |\alpha|} C_{\alpha,\beta} M^\beta + \sum_{|\gamma| \leq |\alpha|-1} \tilde{C}_{\alpha,\gamma} M^\gamma L$$

where $C_{\alpha,\beta}$ and $\tilde{C}_{\alpha,\gamma} \in \Psi^0_{cl}(\omega)$.
Instead of (2.10) we use linearization formula (1.27) of $M^\alpha f(x,u)$ when $|\alpha| = l$. We remark that $Au \in C^\infty(\omega)$ when $A \equiv 0$ because $WF(u) \subset N(\Sigma)$. Thus we get for U^l a linear system of the form

$$LU_i^l + \sum_j A_{i,j} U_j^l = f_i \,,\ f_i \in E_s(\omega)$$

where the linear operators $A_{i,j}$ satisfy properties of the type (1.28), (1.29) in ω. The Proposition 2.4.1, stated and proved at the next paragraph, imply then that $U^l \in E_s(\omega)$ since $U^l \in E_{s-1}(\omega)$ and $U^l_{|\omega^-} \in E_s(\omega^-)$. This means that $u \in H_{s,k}(\omega)$. Since (2.21) is obviously fulfilled when $l = 1$, the proof of Theorem 2.1.2 is complete.

2.4 Linear propagation properties

This paragraph is devoted to the proof of the linear propagation properties used at the end of the proofs of Theorems 2.1.1 and 2.1.2. We have already stated Proposition 2.2.5. More generally one can state the following Proposition.

Proposition 2.4.1 *Proposition 2.2.5 remains valid if the linear system (2.12) is replaced by*

$$LU_i + \sum_j T_{i,j} U_j = f_i \quad (1 \leq i \leq I), \tag{2.22}$$

where the linear operators $T_{i,j}$ fulfil the properties (1.28), (1.29).

More precisely, the following microlocal result holds

Proposition 2.4.2 *For $1 \leq i \leq I$ let U_i be $\mathbb{C}^N$ valued functions belonging to $E_\theta(\omega)$, and satisfying a system of the form (2.22) where $f_i \in E_\sigma(\omega)$, and where $T_{i,j}$ fulfil the properties (1.28), (1.29) with $s > n/2$. We assume that $-s < \theta$, $\sigma \leq s$, $\sigma < \theta + Min(1, s - n/2)$. Then, when $1 \leq i \leq M$, $WF_\sigma(U_i)$ is included in $Char L$, and $\cup_{i=1}^{I} WF_\sigma(U_i)$ is invariant under the flow χ_t defined by the hamiltonian vector field H_p of p. We have denoted by WF_σ the wave front set relative to $E_s(\omega) = H_s^{loc}(\omega)$ or ${}^\omega H_s^{loc}(\omega)$.*

Proof of Proposition 2.4.1 from Proposition 2.4.2.
We first apply Proposition 2.4.2 with $\theta = s - 1$, $\sigma = s - 1 + \rho$, $0 < \rho < Min(1, s - n/2)$. Since $\sigma \leq s$, we get $f_i \in E_\sigma(\omega)$ and $U_{i|\omega^-} \in E_\sigma(\omega^-)$, thus $U_i \in E_\sigma(\omega)$ when $1 \leq i \leq I$. If $s > n/2 + 1$, one can choose $\rho = 1$ and the proof is complete. If $s \leq n/2 + 1$, we apply again Proposition 2.4.2 for $\theta = s - 1 + \rho$ and $\sigma = Min(s, s - 1 + 2\rho)$, and the conclusion is obtained after a finite number of steps.

Proof of the Proposition2.4.2
1) The system (2.22) gives $LU_i = f_i - \sum_j T_{i,j} U_j$, with $f_i \in E_\sigma(\omega)$ and $T_{i,j} U_j \in E_\theta(\omega)$ thanks to (1.28), thus $LU_i \in E_{\sigma-1}(\omega)$ since $\sigma \leq \theta + 1$. The classical inclusion $WF_\sigma(U_i) \subset WF_{\sigma-1}(LU_i) \cup Char L$ shows that $WF_\sigma(U_i) \subset Char L$.

2)Reduction to a global problem in $\mathbb{R}^n$.
Let $\gamma_0 = (x_0, \xi_0) \in WF_\sigma(U) = \cup_{i=1}^{I} WF_\sigma(U_i)$. Our problem is to prove that $\chi_t(\gamma_0) \in WF_\sigma(U)$ when t is in a neighbourhood of 0 in $\mathbb{R}$. Let $\varphi, \psi \in C_0^\infty(\omega)$ such that $\varphi = \psi = 1$ in a neighbourhood of x_0, $\varphi = 1$ in a neighbourhood of the support of ψ. Then $\psi L \varphi U_i = \psi L U_i$ and $\psi T_{i,j} \varphi U_j = \psi T_{i,j} U_j + \psi [T_{i,j}, \varphi] U_j$. But the property (1.29) for $T_{i,j}$ shows that $\psi [T_{i,j}, \varphi] U_j \in E_\sigma(\mathbb{R}^n)$, hence

$$\psi L \varphi U_i + \sum_j \psi T_{i,j} \varphi U_j \in H_\sigma, \quad 1 \leq i \leq I. \tag{2.23}$$

3)Reduction to a first order pseudodifferential system with scalar principal symbol.
Let l be the principal symbol of L, and $\tilde{l}$ the comatrix of l : $\tilde{l}$ is homogeneous of degree $N - 1$ and $\tilde{l} l = pI$ where $p = det l$.
Let $\Lambda = (1 + |D|^2)^{\frac{1-N}{2}} I$. From (2.23) we get

$$\Lambda\, \tilde{l}(x, D)\, \psi\, L\, \varphi\, U_i + \sum_j \Lambda\, \tilde{l}(x, D)\, \psi\, T_{i,j}\, \varphi\, U_i \in E_\sigma, \quad 1 \leq i \leq I.$$

We set $\tilde{U_i} = \varphi U_i$, $\tilde{L} = \Lambda\ \tilde{l}(x,D)\ \psi\ L\ \varphi_1$, $\tilde{T_{i,j}} = \Lambda\ \tilde{l}(x,D)\ \psi\ T_{i,j}\ \varphi_1$ where $\varphi_1 \in C_0^\infty(\omega)$ is such that $\varphi_1 = 1$ in the support of φ.

We have then

$$\tilde{L}\tilde{U_i} + \sum_j \tilde{T_{i,j}}\tilde{U_j} = \tilde{f_i} \ \ (1 \le i \le I), \ \ \tilde{f_i} \in E_\sigma$$

where $\tilde{L}$ is now a global pseudodifferential operators in $I\!R^n$ of degree 1, with principal symbol homogeneous of degree one given by $\psi(x)|\xi|^{1-N}p(x,\xi)I$. Moreover, the operators $\tilde{T_{i,j}}$ satisfy now the properties (1.28), (1.29) globally in $I\!R^n$.

Summing up, Proposition 2.4.2 is a consequence of the following global result.

Proposition 2.4.3 *Let u be a $\mathbb{C}^M$-valued function belonging to $E_\theta(I\!R^n)$ and satisfying a first order linear system of the form*

$$Lu + Tu = f \ , \ f \in E_\sigma(I\!R^n)$$

where L is a $M \times M$ system of classical pseudodifferential operators belonging to $OPS^1(I\!R^n \times I\!R^n)$, with scalar principal symbol $p.I$, p real and homogeneous of degree 1 ; T is a $M \times M$ system of operators $T_{i,j}$ satisfying the following conditions

$$T_{i,j} \ \text{ is continuous in } E_\theta(I\!R^n) \text{ and in } \ E_\sigma(I\!R^n) \tag{2.24}$$

If $A \in OPS^0(I\!R^n \times I\!R^n)$,

$$[A, T_{i,j}] \ \text{ is continuous from } E_\theta(I\!R^n) \ \text{ into } \ E_\sigma(I\!R^n). \tag{2.25}$$

Then $WF_\sigma(u)$ is included in $p^{-1}(0)$ and is invariant under the flow χ_t defined by the hamiltonian vector field H_p of p.

Proof of the Proposition 2.4.3
We denote by $\sigma = \sigma_0 + \sigma_1$ the complete symbol of L, $\sigma_0 = p.I$ for $|\xi| \ge 1$, $\sigma_1 \in S^0(I\!R^n \times I\!R^n)$. It is not hard to prove, thanks to the global estimations of p, that the flow χ_t is globally defined for $t \in I\!R$, and that $\chi_t(y,\eta)$ and its derivatives with respect to (y,η) are bounded for $t \in I\!R$, $y \in I\!R^n$, $\eta \in I\!R^n$, $|\eta| = 1$. Let $\gamma_0 = (x_0, \xi_0) \notin WF_\sigma(u)$. The goal is to prove that $\chi_t(\gamma_0) \notin WF_\sigma(u)$ for all $t \in I\!R$. Let us consider the operator $D_t + L$, where t is a new "time" variable, and where L acts only on the x

variable. Since $\gamma_0 \notin WF_\sigma(u)$, one can find a symbol $\rho_0 \in S^0(\mathbb{R}^n \times \mathbb{R}^n)$ homogeneous of degree 0 for $|\xi| \geq 1$, different from 0 at $(x_0, \lambda\xi_0)$ where $\lambda > 0$ is large, and such that $\rho_0(x, D)u \in E_\sigma$. Let $v = q(t, x, D_x)u$ where $q(t, x, \xi)$ is a classical $M \times M$ matrix valued symbol in $S^0(\mathbb{R}^n \times \mathbb{R}^n)$ with $q \sim \sum_{m \geq 0} q_m$, q_m homogeneous of degree $-m$. We impose the initial condition

$$q(0, x, \xi) = \rho_0(x, \xi).I \tag{2.26}$$

We have $(D_t + L)q(t, x, D_x)u = q(t, x, D_x)h + Ru$ where $h = f - Tu$, $R = [D_t + L, q(t, x, D_x)]$. We shall see that one can choose the symbol q such that the following property is valid:

The operator R is regularizing with respect to the x variable. (2.27)

Indeed, we have $R = r(t, x, D_x)$ with $r(t, x, \xi) = D_t q(t, x, \xi) + b(t, x, \xi)$, $b \in S^0$, $b \sim \sum_{m,\alpha,l} \frac{1}{\alpha!}(\partial_\xi^\alpha \sigma_l D_x^\alpha q_m - \partial_\xi^\alpha q_m D_x^\alpha \sigma_l)$.
The generic term of the last sum is of degree $1 - m - l - |\alpha|$, and vanishes when $l = \alpha = 0$ since $\sigma_0 = p.I$ is a scalar matrix. In particular the term of degree 1 ($m = \alpha = l = 0$) vanishes, thus $b, r \in S^0$ with $r \sim \sum_{k \geq 0} r_k$, r_k homogeneous of degree $-k$ defined by

$$r_k = \partial_t q_k + \sum_{m+l+|\alpha|=k+1; m \leq k} \frac{1}{\alpha!}(\partial_\xi^\alpha \sigma_l D_x^\alpha q_m - \partial_\xi^\alpha q_m D_x^\alpha \sigma_l)$$

Our goal is to prove that q may be choosen in such a way that $r_k = 0$ for all $k \geq 0$ and that (2.26) holds.
Firstly we have

$$r_0 = \partial_t q_0 + \sum_{j=0}^{n}(\partial_{\xi_j}\sigma_0 D_{x_j} q_0 - \partial_{\xi_j} q_0 D_{x_j}\sigma_0) + (\sigma_1 q_0 - q_0 \sigma_1)$$

Since $\sigma_0 = pI$, the condition $r_0 = 0$ is fulfilled if we choose q_0 as a scalar matrix satisfying the transport equation

$$(\partial_t + H_p)q_0 = 0 \tag{2.28}$$

$$\textit{i.e.}\ \frac{d}{dt} q_0(\chi_t(y, \eta)) = 0\ \forall (y, \eta) \in \mathbb{R}^n \times (\mathbb{R}^n \setminus 0).$$

The initial condition (2.26) yields $q_0(0, x, \xi) = \rho_0(x, \xi)$, thus we define $q_0(t, x, \xi) = \rho_0(\chi_t^{-1}(x, \xi))$ for all $(x, \xi) \in \mathbb{R}^n \times (\mathbb{R}^n \setminus 0)$ and $t \in \mathbb{R}$.
When $k \geq 1$, the condition $r_k = 0$ gives

$$(\partial_t + H_p)q_k + (\sigma_1 q_k - q_k \sigma_1) +$$

$$\sum_{m+l+|\alpha|=k+1; m\leq k-1} \frac{1}{\alpha!}(\partial_\xi^\alpha \sigma_l D_x^\alpha q_m - \partial_\xi^\alpha q_m D_x^\alpha \sigma_l) = 0$$

and the initial condition (2.26) gives $q_k(0, x, \xi) = 0$.
Thus the $M \times M$ matrices q_k are defined recursively for $k \geq 1$ by solving Cauchy problems for ordinary differential systems. The global estimations fulfilled by ρ_0 and χ_t easily show that q_k $(k \geq 0)$ satisfies also global estimations allowing to construct $q \in S^0(I\!R^n \times (I\!R^n)$ with $q \sim \sum_{k\geq 0} q_k$.
The function $v = q(t, x, D_x)u$ now satisfies

$$\begin{cases} (D_t + L)v \equiv & q(t, x, D_x)h \mod C(I\!R, E_\sigma) \\ v_{|t=0} = & \rho_0(x, D_x)u \quad \in E_\sigma \end{cases}$$

where $h = f - Tu$; $q(t, x, D_x)f \in C(I\!R, E_\sigma)$; $q(t, x, D_x)Tu \equiv Tq(t, x, D_x)u$ mod. $C(I\!R, E_\sigma)$ thanks to (2.25).

Thus $v \in C(I\!R, E_\theta)$ is a solution of the Cauchy problem

$$\begin{cases} (D_t + L)V = & g - Tv, \quad g \in C(I\!R, E_\sigma) \\ V_{|t=0} & = v_0 \quad \in E_\sigma \end{cases} \tag{2.29}$$

We shall prove that $v \in C(I\!R, E_\sigma)$ thanks to an argument of existence and unicity of the solution of the Cauchy problem (2.29).

1)Existence of a solution V of (2.29) in $C(I\!R, E_\sigma)$.
Starting from $V_0(t, x) = v_0(x)$, we consider the iteration scheme

$$\begin{cases} (D_t + L)V_{l+1} = & g - TV_l \\ V_{l+1}(0, .) = & v_0 \end{cases} \tag{2.30}$$

If $V_l \in C(I\!R, E_\sigma)$, then $g - TV_l \in C(I\!R, E_\sigma)$ thanks to (2.24). Since the principal symbol of L is hermitian, it is well known that the solution V_{l+1} of (2.30) is also in $C(I\!R, E_\sigma)$. Moreover, let $W_l = V_{l+1} - V_l$ where $l \geq 0$. Then $W_{l+1} \in C(I\!R, E_\sigma)$ satisfies the Cauchy problem

$$\begin{cases} (D_t + L)W_{l+1} = & -TW_l \\ W_{l+1}(0, .) = & 0 \end{cases}$$

thus we get the classical energy estimate for $0 \leq t \leq T$, $T > 0$, being fixed:

$$\|W_{l+1}(t, .)\|_\sigma \leq C' \int_0^t \|TW_l(t', .)\|_\sigma dt'.$$

We deduce, using (2.24), that

$$\|W_{l+1}(t,.)\|_\sigma \leq C \int_0^t \|W_l(t',.)\|_\sigma dt'.$$

Denoting $M = Sup_{t\in[0,T]}\|W_0(t,.)\|_\sigma$, we deduce that $\|W_l(t',.)\|_\sigma \leq MC^l t^l/l!$, $l \geq 0$, $t \in [0,T]$.
Therefore V_l is a Cauchy sequence in the space $C([0,T],E_\sigma)$, and converges to a limit V in $C([0,T],E_\sigma)$, which obviously satisfies (2.29).
2)Unicity. Let $W = V - v$. Then $W \in C(I\!R, E_\theta)$ satisfies

$$\begin{cases} (D_t + L)W = & -TW \\ W(0,.) \;\; = & 0 \end{cases}$$

Thus we have now the energy estimate

$$\|W(t,.)\|_\theta \leq C_1 \int_0^t \|W(t',.)\|_\theta dt', \;\; 0 \leq t \leq T$$

which gives $\|W(t',.)\|_\theta \leq M_1 C_1^l t^l/l!$ for $l \geq 0$ and $0 \leq t \leq T$, therefore $W = 0$.
We have proved that $v(t,.) = q(t,x,D_x)u$ belongs to E_σ for any $t \in I\!R$. Since $q(\chi_t(x_0,\lambda\xi_0)) = q_0(x_0,\lambda\xi_0) \neq 0$ when $\lambda > 0$ is large, we obtain that $\chi_t(x_0,\xi_0) \notin WF_\sigma u$ for any $t \in I\!R$, and the proof of Proposition 2.4.3 is complete.

The proof given here is inspired by the proof given by S.Alinhac and P.Gerard [A.G] for the Hörmander's linear propagation Theorem, which corresponds to the scalar pseudodifferential standard case. Slights modifications and simplifications of our proof of Proposition 2.4.2 give the following form of Hörmander's Theorem in the case of systems:

Theorem 2.4.4 *Let ω be an open set in $I\!R^n$; For $1 \leq i \leq I$ let $U_i \in (\mathcal{D}'(\omega))^N$ satisfying a linear system of the form*

$$LU_i + \sum_j T_{ij}U_j = f_i \;\; (1 \leq i \leq I)$$

where L and T_{ij} are $N \times N$ systems of proper classical pseudodifferential operators in ω, of order m, $m-1$ respectively ($m \in I\!R$), $f_i \in E_\sigma(\omega)$, $\sigma \in I\!R$.
We assume that the determinant p of the principal symbol of L is real. Then $\cup_{i=1}^I WF_{\sigma+m-1}(U_i)$ is included in $p^{-1}(\{0\})$ and is invariant under the flow χ_t defined by the hamiltonian vector field of p.

2.5 Remarks

Remark 2.5.1 : *The case of scalar semilinear equation.*
If $L = \sum_{|\alpha|\leq N} a_\alpha(x)D^\alpha$ *is a scalar linear differential operator of order* N *in* $\Omega \subset \mathbb{R}^n$, *we consider a semilinear equation* $Lu = f(x, (\partial^\alpha u)_{|\alpha|\leq N-1})$. *If the principal symbol* p *of* L *is real, we have results similar to Theorems 2.1.1 and 2.1.2 for a solution* $u \in E_s$ *of this equation, under the hypothesis* $s > \frac{n}{2} + N - 1$.

The proof stands on the analogous of the Proposition 2.3.1 (C_j, C_0 are now scalar operators of order $N-1$), which is more simple to prove, since it suffices to establish the two following properties in $\omega \subset\subset \Omega$:

(*) Let $q(x,\xi)$ be a scalar symbol homogeneous of order N. If $q = 0$ on $N(\Sigma_1)$, ..,$N(\Sigma_r)$, then there exists a symbol $c(x,\xi)$ homogeneous of degree 0 such that $q - cp = 0$ on $N(\Sigma_1)$, ..,$N(\Sigma_r)$ and $N(\Gamma)$.

(**) Let $r(x,\xi)$ be a scalar symbol homogeneous of degree N. If $r = 0$ on $N(\Sigma_1)$, ..,$N(\Sigma_r)$ and $N(\Gamma)$, there exists homogeneous symbols c_j of order $N-1$ and generators M_j of $\mathcal{M}(\Sigma \cap \omega)$ such that $r = \sum_j c_j m_j$, where m_j is the principal symbol of M_j.

Remark 2.5.2 *Why the conormal spaces defined with vector fields are not suitable to the proof of Theorem 2.1.2 when* $r \geq 3$.

In fact, the basic commutation property (Proposition 2.3.1 and the analogous Proposition in the scalar case) are not longer true if we replace the pseudodifferential generators M_j of $\mathcal{M}(\Sigma \cap \omega)$ by generators generators Z_j of the Lie Algebra of all vector fields tangent to Σ_1,..,Σ_r. The reason is that these fields may vanish to much at Γ to allow a relation of the type (**). For instance consider in $\mathbb{R}^2$, $\Sigma_1 = \{x = 0\}$, $\Sigma_2 = \{y = 0\}$, $\Sigma_3 = \{x - y = 0\}$. The vector fields tangent to $\Sigma_1, \Sigma_2, \Sigma_3$ are generated by $Z_1 = x\partial_x + y\partial_y$, $Z_2 = x(y-x)\partial_x$. Let $r(x,y,\xi,\eta) = 2(y-x)\xi^2\eta + x\xi^2(\xi+\eta)$. An equality of the form $r = c_1 z_1 + r_2 z_2$ would imply, by considering the first derivatives with respect to x and y for $x = y = 0$ and remarking that z_2 is null at order 2 on Γ:

$$\xi^3 - \xi^2\eta = ic_1\xi\ , \ \ 2\xi^2\eta = ic_1\eta$$

where $c_1 = c_1(0,0,\xi,\eta)$.
This imply the compatibility condition $\xi\eta(\xi+\eta) = 0$, which is violated on $N(\Gamma)$ outside $N(\Sigma_1) \cup N(\Sigma_2) \cup N(\Sigma_3)$.

Chapter 3

Symbolic calculus: propagation; interaction for 2x2 systems.

J.P. VARENNE

3.1 Introduction

We consider a $N \times N$ first order semilinear system

$$Lu = f(x,u) \text{ in } \Omega \subset \mathbb{R}^n \tag{3.1}$$

and we work in the context of conormal solutions studied in chapter one and two.

We are going to precise, in two different cases, the study of the regularity of the solution, phenomena of propagation of some type of regularity, and explain which results can be generalized.

In paragraph 2, we suppose u conormal with respect to Σ, a single simply characteristic hypersurface. In paragraph 3, we suppose $N = 2$, u conormal with respect to two characteristic hypersurfaces Σ_1 and Σ_2, which intersect transversally along a two codimension manifold Γ, assuming that the rays (see 1.1) issued from the points of Γ are transversal to Γ.

We have recalled, in chapter 1.3, the definition of the principal symbol for a conormal distribution. In the case of a single hypersurface Σ, we show that we have for this symbol (which in fact may be slightly modified) the standard transport equation corresponding formally to the linearized

system (see Theorem 3.2.1). In the second case we get for this symbol, well defined outside Γ above each hypersurface $\Sigma_i (i = 1, 2)$, the same transport equation which makes sense across Γ (theorem 3.3.7).
More generally, we will see in chapter 4 and 5 that this result about the principal symbols above $\Sigma_i \setminus \Gamma$ $(i = 1, 2)$ remains true in the general case of interaction described in the Theorem 2.1.2 of J.M. Bony. In the two particular cases studied in this chapter, we can give more informations and develop analysis beyond the principal symbols:
We prove the propagation along the rays of complete asymptotic expansions of convenient type (see Theorem 3.2.7, 3.3.8). This kind of result (and similar results for the Cauchy problem and for boundary problems) was proved in the case of classical (with or without logarithms) conormal distributions in Melrose R. B. and Ritter N. [MR], Rauch J. and Reed M. [$RR4$], [$RR5$], Piriou A. [P], Nadir B. and Piriou A. [$NP1$], [$NP2$], Nadir B. and Varenne J. P. [$NV2$], Joshi M.S. [J].
The presentation given here is new, and implies all the different cases already known. In the general situation of interaction described in the Theorem 2.1.2, the analysis beyond the principal symbols is an open problem when the size N of the system is larger than 2. Moreover, some examples are studied in chapter 5.
We develop here only the case where the solution of (3.1) is a $I\!R^n$ valued solution, to give a more pleasant presentation of the proofs, but the complex case can be treated in a similar way.

3.2 The case of a single conormal wave.

We consider a first order semilinear system:

$$Lu = f(x, u) \tag{3.2}$$

with $x \in \Omega$, open set in $I\!R^n$, L linear differential system. The coefficients of L are $N \times N$ smooth real matrices. The $I\!R^N$ valued function f is smooth.
Let Σ be a closed simply characteristic hypersurface. We consider a $I\!R^N$ valued solution u of (3.2) belonging to the conormal - Besov based algebra $H_s(\Sigma)$ with $s > \frac{1}{2}$ (see the Definition 1.2.1 and Remark 1.2.14). We don't discuss here the existence of such a solution, relying on the general existence theorems in the Sobolev and Besov spaces and on the Bony's theorems in the case $s > \frac{n}{2}$, and also on results of Ritter N. [R], Métivier G. [$M1$], [$M2$], Beals M. and Métivier G. [BM] in other cases. We note,

in this part, $I^\mu \equiv I^\mu(\Sigma) \equiv H_s(\Sigma)$ for $\mu = -s - \frac{1}{2}$ (see (1.32)).
Recall (see Lemma 2.2.2) that dim $(kerl(x,\xi)) = 1$ when $(x,\xi) \in N(\Sigma)$, where l is the principal symbol of L. We consider locally a smooth non null vector $e_1(x) \in I\!R^N$ such that $e_1(x) \in ker(l(x,\xi))$ if $(x,\xi) \in N(\Sigma)$ and a smooth non null vector $e_1'(x) \in I\!R^N$ such that $e_1'(x) \in ker({}^t l(x,\xi))$ if $(x,\xi) \in N(\Sigma)$. We define the first order scalar differential operator $L_{1,1}$ by the formula

$$L_{1,1}(v) = < e_1', L(ve_1) > .$$

Let H_1 be the Hamiltonian field and h_1 the subprincipal symbol of $L_{1,1}$. We recall (see Proposition 2.2.6) that H_1 is proportional to the Hamiltonian field of L on $N(\Sigma)$.
The principal symbol of u has been defined in (1.34). To take in account the non linearity of $f(x,u)$, we give a modification of the definition:
For $\delta \in]0,1]$, we define the δ- principal symbol of u by replacing in (1.34) $I^{\mu-1}$ and $S^{\mu-\frac{1}{2}}$ by $I^{\mu-\delta}$ and $S^{\mu+\frac{1}{2}-\delta}$. We note:

$$\delta = \delta(\mu) = \begin{cases} min(1, -(\mu+1)) & \text{if } \mu \neq -2 \\ \text{any number in }]0,1[& \text{if } \mu = -2 \end{cases} \tag{3.3}$$

Theorem 3.2.1 *With our hypothesis and notations, let u a solution of (2.1) in $I^\mu(\Sigma)$, $\mu < -1$. Let $\mu' \leq \mu$. If $u \in I^{\mu'}(\Sigma)$ close to a point x_0 of Σ, then $u \in I^{\mu'}(\Sigma)$ close to the ray C of Σ through x_0.*
Let $\delta' \equiv \delta(\mu')$. The δ'-principal symbol of u in $I^{\mu'}(\Sigma)$ is represented by $\sigma.e_1$, where the half density σ satisfies on $N(\Sigma)_{|C}$ the transport equation:

$$(\frac{1}{i}\mathcal{L}_{H_1} + h_1)\sigma - f_1'\sigma = 0 \tag{3.4}$$

$\mathcal{L}_{H_1}$ is the lie-derivate with respect to H_1, f_1' is the function on Σ defined by:

$$f_1' = < e_1', f_u'(.,u)e_1 >_{|\Sigma} .$$

To avoid repetitions, we will give the proof after the one of the next Theorem, which states a general property of propagation along the rays of some conormal class of solutions.
We consider a class $E \equiv E(\Sigma)$ of conormal distributions statisfying the following properties, (3.5), (3.6), $(P_{3.i}), 1 \leq i \leq 4$:

$$E \subset I^{\mu_0} \quad \text{for some } \mu_0 < -1. \tag{3.5}$$

$$C^\infty \subset E. \tag{3.6}$$

($\mathbf{P_{3.1}}$) Multiplication, tangential derivation let E stable.
($\mathbf{P_{3.2}}$) Regularity: Let Z be a transversal field to Σ, $f \in E$, then:

$$Zu = f, \ \ u \in C^\infty \text{ outside } \Sigma \Longrightarrow u \in E.$$

($\mathbf{P_{3.3}}$) Propagation: Let Z a tangential field to Σ, $f \in E$, $u \in \mathcal{D}'$. Suppose f and u vanish near $x_0 \in \Sigma$. If $Zu = f$, then $u \in E$ near each point of the trajectory of Z on Σ issued from x_0.

($\mathbf{P_{3.4}}$) Approximation: Let $v \in \mathcal{D}'$ such that: $\forall \mu \in I\!R$, $v = u_\mu + r_\mu$, with $u_\mu \in E$, $r_\mu \in I^\mu$. Then $v \in E$.

Remark 3.2.2 *If $f \in I^\mu$, $(P_{3.2})$ can be precised :*
$Zu = f$, $u \in C^\infty$ *outside* $\Sigma \Rightarrow u \in I^{\mu-1}$.

We will need the following spaces:
For $\mu < -1$, we define $k(\mu) \in I\!N$ by :

$$-2 \leq \mu + k(\mu) < -1 \tag{3.7}$$

Definition 3.2.3 *For $\mu < -1$, let $\tilde{I}^\mu \equiv \tilde{I}^\mu(\Sigma)$ the subset of $u \in I^\mu$ whose derivatives of order less than or equal to $k(\mu)$ vanish on Σ.*

For instance, if $\Sigma = \{x_1 = 0\}$, $\Re(d) > 0$, $k \in I\!N$, the distribution $\mathcal{H}(x_1)x_1^d(\ln x_1)^k$ belongs to $\tilde{I}^\mu(\Sigma)$ if $\mu = -\Re(d) - 1 + \epsilon$, $\forall \epsilon > 0$.
Recall (see A. Piriou [*P*]) some multiplicative properties of these spaces:

Proposition 3.2.4 *(i) Let $u_j \in \tilde{I}^{\mu_j}$, $j = 1, 2$, $\mu_j < -1$. Then $u_1 u_2 \in \tilde{I}^{\mu_1+\mu_2+1}$ if $\mu_j \notin Z\!\!Z$, $\tilde{I}^{\mu_1+\mu_2+1+\epsilon}$, $(\forall \epsilon > 0)$ if μ_1 or $\mu_2 \in Z\!\!Z$.*
(ii) Let $u_1 \in \tilde{I}^{\mu_1}$, $u_2 \in I^{\mu_2}$, with $\mu_1, \mu_2 < -1$. Then $u_1 u_2 \in \tilde{I}^{\mu_1}$.

We give now some examples for E .

Examples 3.2.5

(1)$E = I^\mu$ with $\mu < -1$.

(2) Classical conormal distributions (see the last paragraph of Chapter one).

Let $D \subset \mathbb{C}$ such that :

$$D \subset \{\Re(d) > 0\} \cup \{0\}$$

$$\mathbb{N} \subset D$$

$$\forall M \in \mathbb{R}, \ \{d \in D / \Re(d) \leq M\} \text{ is finite.}$$

A conormal distribution u with respect to Σ is classical if near each point of Σ , for local coordinates such that : $\Sigma = \{x_1 = 0\}$, $x = (x_1, x')$,

$$u \sim \sum_{d \in D} \sigma_d^{\mp}(x') | x_1 |^d \tag{3.8}$$

for $\mp x_1 > 0$, $\sigma_d^{\mp} \in C^\infty$, in the sense that:

$$u - \sum_{\Re(d) < M} \sigma_d^{\mp}(x') | x_1 |^d \in \tilde{I}^{-1-M}$$

To insure non linear properties (as $P_{3,1}$), we suppose also :

$$D + D \subset D \text{ and thus } D + \mathbb{N} \subset D,$$

$$\bar{D} = D.$$

With all these conditions on D, we can take for E the set of classical conormal distributions, with the condition $\sigma_0^+(x') = \sigma_0^-(x')$ for (3.5). In fact, we could study the discontinuous solutions by a similar method, introducing distributions which are restrictions of distributions of $E(\Sigma)$ on each side of Σ (see for example, Rauch-Reed [$RR5$]) .

(3) Piecewise smooth functions.
It is the particular case of the example (2) corresponding to $D = \mathbb{N}$.

(4) Classical with logarithms conormal distributions (see the last paragraph of Chapter 1).
We define E in the same way, replacing in (3.8) homogeneous terms by multihomogeneous terms:

$$u \sim \sum_{d \in D} p_d^{\mp}(x', \ln | x_1 |) | x_1 |^d \tag{3.9}$$

for $\mp x_1 > 0$, where $p_d^{\mp}$ is polynomial in $\ln|x_1|$, with smooth coefficients with respect to x', in the sense that:

$$u - \sum_{\Re(d)<M} p_d^{\mp}(x', \ln|x_1|)|x_1|^d \in \cap_{\epsilon>0} I^{-1-M+\epsilon}$$

Note that, in the examples above, the condition $(P_{3.4})$ result from the unicity of the asymtotic developments (3.8),(3.9).

Proposition 3.2.6 *Let $f(x,U)$ a smooth function on $\Omega \times \mathbb{C}^r$. If $u \in E^r$, then $f(x,u) \in E$.*

Proof: For $i = 1, ..., r$, we can write $u_i = \phi_i + v_i$, with $\phi_i \in C^\infty$, $v_i \in E \cap \tilde{I}^{\mu_o}$. In local coordinates, where $\Sigma = \{x_1 = 0\}$, $x = (x_1, x')$, we can take :

$$\phi_i(x) = \sum_{0 \le j \le k(\mu_o)} \frac{x_1^j}{j!} \partial_1^j u_i(0, x')$$

We apply Taylor's formula to $f(x,u)$ at $(\phi_1, .., \phi_r)$. With (3.6) ,$(P_{3.1})$, and repeated applications of proposition 3.2.4, we find :

$$\forall \mu, \ f(x,u) \in E + I^\mu.$$

We conclude that $f(x,u) \in E$ using $(P_{3.4})$.

Theorem 3.2.7 *With the hypothesis of theorem 3.2.1, let $u \in I^{\mu_o}$ a $\mathbb{R}^N$-valued solution of (3.1) with $\mu_o < -1$. If $u \in E$ near $x_o \in \Sigma$, then $u \in E$ near each point of the ray γ on Σ issued from x_o.*

Proof: Let $A = \{x \in \gamma \ / \ u \in E \text{ near } x\}$. We prove that A is closed in γ. Let y in the closure of A. Near y, using the reduction of the lemma (2.2.4), (3.1) becomes:

For $\Sigma = \{x_1 = 0\}$, $x = (x_1, x') = (x_1, x'', x_n)$, $y = (x_1, x'', 0)$,

$$\partial_n u_1 + \sum_{j=2,..,n} A_{1,j}(x)\partial_j v = f_1(x,u) \tag{3.10}$$

$$\partial_1 v + \sum_{j=2,..,n} B_j(x)\partial_j u = F(x,u) \tag{3.11}$$

with $u = (u_1, v)$ and $A_{1,j}$, $B_j \in C^\infty$, $F(x,u) = (f_2, .., f_N)(x,u)$.

The operators $T_1 = \sum_{j\geq 2} A_{1,j}(x)\partial_j$, $T_2 = \sum_{j\geq 2} B_j(x)\partial_j$ let I^μ stable. We proceed by induction. Suppose that : For $\mu \leq \mu_0$,

$$(H_\mu)\begin{cases} u_1 &= \phi + U + R \\ v &= \psi + V + r \end{cases}$$

with $(\phi, \psi) \in C^\infty$, $U$and $V \in E \cap \tilde{I}^{\mu_o}$, $R \in \tilde{I}^{\mu_o}$, $r \in \tilde{I}^{\mu_o - 1}$.

Note that (H_μ) implies : $u \in E + I^\mu$.
Note also that, since $E \subset I^{\mu_o}$ and the restriction of I^μ to Σ is C^∞, we have: $E + I^\mu \subset E \cap \tilde{I}^{\mu_o} + I^\mu$.

The hypothesis (H_μ) is true for $\mu = \mu_o$, with $U = V = 0$, since we have the :

Lemma 3.2.8 *(Polarization)* $v \in I^{\mu_o - 1}$.

Proof of lemma: The equation (3.11) gives, using proposition 1.2.6: $\partial_1 v \in I^{\mu_0}$, so, by integrating, the lemma follows easily.
Suppose now that (H_μ) is true. We will prove $(H_{\mu-\delta})$ for $\delta = \delta(\mu_o)$ (see (3.3)).

First step: We prove $(H_{\mu-\delta})$ for u_1. Using (3.10), we have:

$$\partial_n u_1 = f_1(x,u) - T_1(v). \tag{3.12}$$

Using $(P_{3.1})$, we obtain:

$$T_1(v) \in E + I^{\mu-1}. \tag{3.13}$$

By Taylor's formula, we have, using (H_μ) :

$$f_1(u) \equiv f_1(x,u) = f_1(\phi + U, \psi + V) + \frac{\partial f_1}{\partial u_1}(\phi + U, \psi + V)R +$$

$$\frac{\partial f_1}{\partial v}(\phi + U, \psi + V)r + T, \tag{3.14}$$

where T is a sum of terms as: $F(\phi, \psi, U, V, R, r)R^2$, with $F \in C^\infty$. By C^∞stability of I^{μ_o}, proposition 3.2.4, we find:

$$T \in I^{\mu-\delta}. \tag{3.15}$$

By Taylor's formula, we find in the same way :

$$\frac{\partial f_1}{\partial u_1}(\phi+U, \psi+V)R \equiv \frac{\partial f_1}{\partial u_1}(\phi, \psi)R + I^{\mu-\delta}. \tag{3.16}$$

Proposition 3.2.4 and (H_μ) give :

$$\frac{\partial f_1}{\partial v}(\phi+U, \psi+V)r \in I^{\mu-\delta}. \tag{3.17}$$

Proposition 3.2.6 gives:

$$f_1(\phi+U, \psi+V) \in E. \tag{3.18}$$

Using (3.14) to (3.18), we get:

$$f_1(u) = \frac{\partial f_1}{\partial u_1}(\phi, \psi)R \quad [E + I^{\mu-\delta}]. \tag{3.19}$$

Using (3.13), (3.12) gives:

$$\partial_n R - \frac{\partial f_1}{\partial u_1}(\phi, \psi)R = g \in E + I^{\mu-\delta}. \tag{3.20}$$

By hypothesis, we know that: $\exists \epsilon' > 0 / \; u \in E$ for $x_n < -\epsilon'$. Then:

$$R \in E \quad \text{for} \quad x_n < -\epsilon'. \tag{3.21}$$

We introduce a cut off function $Z(x_n)$, C^∞ so that:

For $0 < \epsilon < \epsilon' < \epsilon''$, we have : $Z(x_n) = 1$ for $x_n \leq -\epsilon'$,

$$Z(x_n) = 0 \text{ for } x_n \leq -\epsilon''. \tag{3.22}$$

Then $R \;\equiv\; RZ$ near y.
Let $Y = RZ$, (3.20) gives:

$$\partial_n(y) - \frac{\partial f_1}{\partial u_1}(\phi, \psi)y = h,$$

with $h = Zg + Z^{'}R \in E + I^{\mu-\delta}$, using (3.21), (3.22).
We find :

$$\partial_n(y) - \frac{\partial f_1}{\partial u_1}(\phi, \psi)y \in E + I^{\mu-\delta}$$

and $y \equiv 0$ for $x_n \leq -\epsilon^{''}$

Using $(P_{3.3})$, remark 3.2.2, we find: $y \in E + I^{\mu-\delta}$, by an explicit integration. So, $R \in E + I^{\mu-\delta}$ near y, and we have $(H_{\mu-\delta})$ for u_1.

Second step: We prove $(H_{\mu-\delta})$ for v.

Using (3.11), we have:

$$\partial_1 v \quad = \quad F(u) - T_2(u). \tag{3.23}$$

The first step shows:

$$u, \;\; T_2(u) \in E + I^{\mu-\delta}. \tag{3.24}$$

The same arguments as those we have used in proposition 3.2.6 prove

Lemma 3.2.9 *For all $\mu < -1$, $E + I^{\mu}$ is stable by smooth function.*

We get, using lemma 3.2.9:

$$F(u) \in E + I^{\mu}. \tag{3.25}$$

Using (3.23) to (3.25) we find:

$$\partial_1 v \in E + I^{\mu-\delta} \;,$$

so $(P_{3.2})$, (3.5), remark 3.2.2 give:

$$v \in E + I^{\mu-1-\delta} \subset E \cap \tilde{I}^{\mu_0} + I^{\mu-1-\delta}.$$

We have proved the second step, and $(H_{\mu-\delta})$.
So, $u \in E + I^{\mu-\delta}$, and by iteration, $\forall \mu^{'}$, $u \in E + I^{\mu^{'}}$, and $(P_{3.4})$ gives: $u \in E$. We have proved theorem 3.2.7.

Proof of Theorem 3.2.1: The first part is the particular case of Theorem 3.2.7 for $E \equiv I^{\mu^{'}}$. The proof of the second part follows from the proof of Theorem 3.2.7: Close to C, we have the property (H_{μ}) with $\mu = \mu^{'}$, $U = V = 0$, and we obtain (3.20) with $g \in I^{\mu^{'}-\delta^{'}}$ which shows easily (3.4) in our local coordinates.

3.3 The case of transversal interaction for 2×2 systems

We consider now a system (3.2) with size $N = 2$. Let Σ_1, Σ_2 two characteristic hypersurfaces intersecting transversally along a 2 codimension manifold Γ.
We suppose: The rays on Σ_1, Σ_2 issued from points of Γ are transversal to Γ.
Let u a $\mathbb{R}^2$-valued solution of (3.2) belonging to the conormal-based Besov algebra $H_s(\Sigma_1 \cup \Sigma_2)$ with $s > 1$ (see Definition 1.2.1) and the considerations at the beginning of paragraph 2 about the existence of such solution).
We introduce new conormal distribution spaces which are, in some sense, the completion of the algebraic tensor product of two spaces $E(\Sigma_1)$ and $E(\Sigma_2)$.

Definition 3.3.1 *For $\mu, \nu \in \mathbb{R}$, we note $J^{\mu,\nu} \equiv J^{\mu,\nu}(\Omega)$ the space of distributions u in Ω so that:*
- u is C^∞ outside $\Sigma_1 \cup \Sigma_2$.
- $u \in I^\mu(\Sigma_1)$ near each point of $\Sigma_1 \setminus \Gamma$, and $u \in I^\nu(\Sigma_2)$ near each point of $\Sigma_2 \setminus \Gamma$.
- Near each point of Γ, for local coordinates such that: $x = (x_1, x_2, x")$, $\Sigma_i = \{x_i = 0\}$, $(i = 1, 2)$, one has:

$$u(x) = \int e^{ix_1\xi_1 + ix_2\xi_2} a(x", \xi_1, \xi_2) d\xi_1 d\xi_2, \tag{3.26}$$

with $a \in C^\infty(\mathbb{R}^{n-2} \times \mathbb{R} \times \mathbb{R})$ satisfying: $\forall (\alpha_1, \alpha_2, \beta), \exists C \geq 0$ with,

$$| \partial_{x"}^\beta \partial_1^{\alpha_1} \partial_2^{\alpha_2} a(x", \xi_1, \xi_2) | \leq C(1+ | \xi_1 |)^{\mu-\alpha_1}(1+ | \xi_2 |)^{\nu-\alpha_2} \tag{3.27}$$

(we note ∂_i the derivative with respect to ξ_i).

Definition 3.3.2 *Let $\mu, \nu < -1$. We denote by $J_1^{\mu,\nu}$ the subset of $u \in J^{\mu,\nu}$ so that: $Z_1..Z_j u = 0$ on Σ_1, $\forall Z_j$, tangential field to Σ_2, for $j \leq k(\mu)$, where $k(\mu)$ is defined in (3.7).*

We define symmetrically the space $J_2^{\mu,\nu}$ and we put $J_{1,2}^{\mu,\nu} = J_1^{\mu,\nu} \cap J_2^{\mu,\nu}$.

For example, we have: $\tilde{I}^\mu(\Sigma_1) \otimes I^\nu(\Sigma_2) \subset J_1^{\mu,\nu}$.
We have an analogous to the proposition 3.2.4:

Proposition 3.3.3 *(see B. Nadir and A. Piriou [NP2])*
Let $\mu, \nu, \mu', \nu' < -1$.
(i) If $u \in J_1^{\mu,\nu}$, $u' \in J_1^{\mu',\nu}$, then $uu' \in J_1^{\mu+\mu'+1,\nu}$ if $\mu, \mu' \notin \mathbb{Z}$.

(ii)If $u \in J_{1,2}^{\mu,\nu}$, $u' \in J_{1,2}^{\mu',\nu'}$, then $uu' \in J_{1,2}^{\mu+\mu'+1,\nu+\nu'+1}$ if $\mu, \mu', \nu, \nu' \notin \mathbb{Z}$.

(iii) If $u \in J^{\mu,\nu}$, $u' \in J_1^{\mu',\nu}$, then $uu' \in J_1^{\mu',\nu}$

If $u \in J^{\mu,\nu}$, $u' \in J_{1,2}^{\mu',\nu'}$, then $uu' \in J_{1,2}^{\mu',\nu'}$

If $u \in J_1^{\mu,\nu}$, $u' \in J_2^{\mu',\nu'}$, then $uu' \in J_{1,2}^{\mu,\nu'}$

Proposition 3.3.4 *(see B. Nadir and A. Piriou [NP2] th. 2.7)*
Let $f(x,U)$ a smooth function on $\mathbb{R} \times \mathbb{C}^r$. If $u \in [J^{\mu,\nu}]^r$ with $\mu, \nu < -1$, then $f(x,u) \in J^{\mu,\nu}$.

We note $E \equiv E(\Sigma_1)$, $E' \equiv E(\Sigma_2)$.

Definition 3.3.5 *We set $P(E,E') = \{u \in \mathcal{D}'(\Omega)/\forall\mu, \forall\mu', u \in E \otimes E' + E \otimes I^{\mu'}(\Sigma_2) + I^{\mu}(\Sigma_1) \otimes E' + J^{\mu,\mu'}\}$, where $\otimes$ denotes the algebraic tensor product.*

Near each point of $\Sigma_1 \setminus \Gamma$, for example, thanks to $(P_{3.4})$ we have:

$$P(E,E') \equiv E.$$

We add to $P_{3.i}$, $i = 1,..,4$, the hypothesis:

$(\mathbf{P_{3.5}})$ Let Z a transversal field to Σ_1, tangential to Σ_2. Suppose that f and u vanish near $x_0 \in \Sigma_2$. Suppose $Zu = f$.
If $f \in E \otimes E'$, then $u \in E \otimes E'$
If $f \in E \otimes I^{\nu}(\Sigma_2)$, then $u \in E \otimes I^{\nu}(\Sigma_2)$
If $f \in I^{\mu}(\Sigma_1) \otimes E'$, then $u \in I^{\mu-1}(\Sigma_1) \otimes E'$.
We suppose also that the similar property is true if we permute Σ_1 and Σ_2 in the hypothesis.

Remark 3.3.6 *The property $(P_{3.5})$ is true for the classe of conormal distributions defined in examples 3.2.5. Note also that, in the first case of $(P_{3.5})$, if $f \in J^{\mu,\nu}$, then $u \in J^{\mu-1,\nu}$.*

We use the same notations as the reduction 2.3.2, in the case $r = N = 2$. Recall we have defined a basis $(e_1(x), e_2(x))$ of $I\!R^2$ introducing $\ker l(x,\xi)$ restricted to $N(\Sigma_j)$, ($j = 1,2$), and a similar basis $(e_1^{'}(x), e_2^{'}(x))$ introducing $\ker({}^t l(x,\xi))$. We have also defined in (2.15) the operators $L_{j,k}$ for $j,k \leq 2$. For $j = 1,2$, we note H_j the Hamiltonian field of $L_{j,j}$ and h_j the subprincipal symbol of $L_{j,j}$.

Theorem 3.3.7 *Let $x_0 \in \Gamma$, C_1, C_2, the two rays on Σ_1, Σ_2, issued from a point $x_0 \in \Gamma$. We note C_1^- (respectively C_2^-) a connected component of $C_1 \setminus \{x_0\}$ (respectively $C_2 \setminus \{x_0\}$). Let $u \in H_s(\Sigma_1 \cup \Sigma_2)$ a $I\!R^2$-valued solution of (3.2) with $N = 2$, $s > 1$.*
Suppose that $u \in I^{\mu}(\Sigma_1)$ near a point of C_1^-, $u \in I^{\mu^{'}}(\Sigma_2)$ near a point of C_2^-, with $\mu^{'} \leq \mu \leq -s - \frac{1}{2}$. Then $u \in J^{\mu,\mu^{'}}$ near x_0, and thus, $u \in I^{\mu}(\Sigma_1)$ near $C_1 \setminus \{x_0\}$, $u \in I^{\mu^{'}}(\Sigma_2)$ near $C_2 \setminus \{x_0\}$.
Let $\delta = \delta(\mu)$ (see (3.3)). For $j = 1,2$, the δ-principal symbol of u on $N(\Sigma_j)$ near points of $C_j \setminus \{x_0\}$ is represented by $\sigma_j e_j$, where we have, for the half density σ_j, the transport equation through Γ

$$(\frac{1}{i}\mathcal{L}_{H_j} + h_j)\sigma_j = f_j^{'}\sigma_j$$

$\mathcal{L}_{H_j}$ is the Lie-derivate with respect to H_j, $f_j^{'}$ is the function on Σ_j gived by:

$$f_j^{'} = < e_j^{'}, f_u^{'}(.,u)e_j >_{|\Sigma_j} .$$

The proof will be given after the proof of the next theorem.

Theorem 3.3.8 *With notations of theorem 3.3.7, let $u \in H_s(\Sigma_1 \cup \Sigma_2)$ a $I\!R^2$-valued solution of (3.2) with $N = 2$, $s > 1$. Suppose that $u \in E(\Sigma_1)$ near C_1^- with $\mu_0 = -s - \frac{1}{2}$, $u \in E(\Sigma_2)$ near C_2^- with $\mu_0^{'} \leq \mu_0$, where μ_0, $\mu_0^{'}$ satisfy (3.5) for $E(\Sigma_1)$, $E(\Sigma_2)$, respectively. Then $u \in P(E, E^{'})$ near x_0. So, $u \in E(\Sigma_1)$ near $C_1 \setminus \{x_0\}$, $u \in E(\Sigma_2)$ near $C_2 \setminus \{x_0\}$.*

Proof of Theorem3.3.8:
We work near x_0, in suitable local coordinates (see Reduction 2.3.2 and Proposition 2.3.3), such that: $\Sigma_i = \{x_i = 0\}$, $x_0 = 0$, (3.2) becomes:

$$L_{1,1}(u_1) + L_{1,2}(u_2) = f_1(x,u) \tag{3.28}$$

$$L_{2,1}(u_1) + L_{2,2}(u_2) = f_2(x,u) \tag{3.29}$$

For $x_i = 0$, $L_{i,i}$ is proportional to the bicharacteristic field on Σ_i, and, for $i \neq j$, $L_{i,j}$ is a tangential field to (Σ_i, Σ_j).

We note: $E_0 \equiv E \cap \tilde{I}^{\mu_0}(\Sigma_1)$, $E_0^{'} \equiv E^{'} \cap \tilde{I}^{\mu_0^{'}}(\Sigma_2)$, $F_{\mu,\mu'} \equiv E_0 \otimes E_0^{'} + E_0 \otimes I^{\mu^{'}}(\Sigma_2) + I^{\mu}(\Sigma_1) \otimes E_0^{'} + J^{\mu,\mu^{'}}$.
We have $P(E, E^{'}) \equiv \{u \in J^{\mu_0,\mu_0^{'}} \,/\, u \in F_{\mu,\mu'},\ \forall \mu, \mu^{'}\}$.

We proceed by a double induction. We make the hypothesis: For $\mu \leq \mu_0$, $\mu^{'} \leq \mu_0^{'}$,

$$(H_{\mu,\mu'}) \left\{ \begin{array}{lcl} u_1 & = & u_{1,0,0} + v_{1,0} + w_{1,0} + r_1 \;\in\; F_{\mu,\mu'-1}, \\ u_2 & = & u_{2,0,0} + v_{2,0} + w_{2,0} + r_2 \;\in\; F_{\mu-1,\mu'}, \end{array} \right.$$

with $u_{i,0,0} \in E_0 \otimes E_0^{'}$, $v_{1,0} \in E_0 \otimes I^{\mu^{'}-1}(\Sigma_2)$, $w_{1,0} \in I^{\mu}(\Sigma_1) \otimes E_0^{'}$, $r_1 \in J^{\mu,\mu^{'}-1}$, $v_{2,0} \in E_0 \otimes I^{\mu^{'}}(\Sigma_2)$, $w_{2,0} \in I^{\mu-1}(\Sigma_1) \otimes E_0^{'}$, $r_2 \in J^{\mu-1,\mu^{'}}$.

We suppose that, for $\mu = \mu_0$, $\mu^{'} = \mu_0^{'}$: $u_{i,0,0} \equiv v_{i,0} \equiv w_{i,0} \equiv 0$ for $i = 1, 2$.

The case $\mu = \mu_0$, $\mu^{'} = \mu_0^{'}$ comes from:

Lemma 3.3.9 *(Polarization)* $u \in J^{\mu_0,\mu_0^{'}}$, $u_1 \in J^{\mu_0,\mu_0^{'}-1}$, $u_2 \in J^{\mu_0-1,\mu_0^{'}}$.

This lemma is a direct consequence of the general polarization we will see in Chapter 4 (Proposition 4.3.1).

We prove that $(H_{\mu,\mu'})$ implies $(H_{\mu-\delta,\mu'})$, with $\delta = \delta(\mu_0)$. So, by symmetry, the double induction will be proved.

We note $f_1(u) \equiv f_1(x, u)$.
First step: We prove $(H_{\mu-\delta,\mu'})$ for u_1. Using (3.28), we have:

$$L_{1,1}(u_1) \;=\; f_1(u) - L_{1,2}(u_2) \tag{3.30}$$

Let, for $i = 1, 2$: $v_i = u_{i,0,0} + v_{i,0}$, $w_i^{'} = w_{i,0} + r_i$.

We write:

$$w_i^{'} \;=\; \phi_i + w_i, \tag{3.31}$$

with $\phi_i = \sum_{0\leq j\leq k_i} \frac{x_1^j}{j!}\partial_1^j w_i^{'}(0,x^{'})$, for $k_1 = k(\mu)$, $k_2 = k(\mu-1)$ (see(3.7)). $\phi_1 \in E^{'} + I^{\mu^{'}-1}(\Sigma_2)$, $\phi_2 \in E^{'} + I^{\mu^{'}}(\Sigma_2)$, $w_1 \in E_1^{\mu,0} + J_1^{\mu,\mu^{'}-1}$, $w_2 \in E_1^{\mu-1,0} + J_1^{\mu-1,\mu^{'}}$, where $E_1^{\mu,0}$ denote the subset of $I^{\mu}(\Sigma) \otimes E_0^{'}$ whose derivatives of order less than or equal to $(k(\mu)+1)$ vanish on Σ_1.
$E_2^{0,\mu}$ is defined in a symmetric way.
We write:

$$u_i \quad = \quad \phi_i + v_i + w_i. \tag{3.32}$$

We linearize $f_1(u)$ with these decompositions:

$$f_1(u) = f_1(\phi_1+v_1,\phi_2+v_2) + \sum_{k=1}^{2} \frac{\partial f_1}{\partial u_k}(\phi_1+v_1,\phi_2+v_2)w_k +$$

$$\sum_{i+j=2} S_{i,j}(\phi_1+v_1,\phi_2+v_2,w_1,w_2)w_i w_j, \tag{3.33}$$

with $S_{i,j} \in C^{\infty}$.
For the study of the last term, we write, as in (3.31) (but now we take the restrictions of the different terms to Σ_2): For $i = 1,2$,

$$\phi_i \quad = \quad \phi_i^{'} + \phi_i^{"}, \tag{3.34}$$

with $\phi_i^{'} \in C^{\infty}$, $\phi_i^{"} \in E_0^{'} + \tilde{I}^{\mu^{'}}(\Sigma_2)$.

$$v_i \quad = \quad v_i^{'} + v_i^{"}, \tag{3.35}$$

with $v_i^{'} \in E_0$, $v_i^{"} \in E^{0,0} + E_2^{0,\mu^{'}}$.

$$w_i \quad = \quad w_i^{'} + w_i^{"}, \tag{3.36}$$

with $w_i^{'} \in \tilde{I}^{\mu}(\Sigma_1)$, $w_i^{"} \in E_1^{\mu,0} + J_{1,2}^{\mu,\mu^{'}}$.

For the study of terms $S_{i,j}(\phi_1+v_1,\phi_2+v_2,w_1,w_2)w_i w_j$, we apply Taylor's formula at $(\phi_1^{'}+v_1^{'},\phi_2^{'}+v_2^{'},w_1^{'},w_2^{'})$ to $S_{i,j}$. Using lemma 3.2.9, we know that

$$S_{i,j}(\phi_1^{'}+v_1^{'},\phi_2^{'}+v_2^{'},w_1^{'},w_2^{'}) \in E + I^{\mu}(\Sigma_1).$$

Then, using proposition 3.2.4 and (3.26), (3.5), $(P_{3.1})$, we find:

$$S_{i,j}(\phi_1 + v_1, \phi_2 + v_2, w_1, w_2)w_i w_j \in T_{\mu-\delta,\mu'} \equiv E_1^{\mu-\delta,0} + J_1^{\mu-\delta,\mu'} \quad (3.37)$$

Using similar arguments, we get:

$$\frac{\partial f_1}{\partial u_i}(\phi_1 + v_1, \phi_2 + v_2)w_i \equiv \frac{\partial f_1}{\partial u_i}(\phi_1, \phi_2)w_i \ [T_{\mu-\delta,\mu'}], \quad (3.38)$$

for $i = 1, 2$,

$$\frac{\partial f_1}{\partial u_2}(\phi_1, \phi_2)w_2 \in T_{\mu-\delta,\mu'}, \quad (3.39)$$

$$f_1(\phi_1 + v_1, \phi_2 + v_2) \in F_{\nu,\mu'}, \ \forall \nu \quad (3.40)$$

Using (3.33), (3.37) to (3.40), lemma 3.2.9, we find:

$$f_1(u) = \frac{\partial f_1}{\partial u_1}(\phi_1, \phi_2)w_1 \ [F_{\mu-\delta,\mu'}], \quad (3.41)$$

with $\frac{\partial f_1}{\partial u_1}(\phi_1, \phi_2) \in E' + I^{\mu'}(\Sigma_2)$.
Since $L_{1,2}$ is a tangential field to (Σ_1, Σ_2), it preserves the spaces $I^{\mu}(\Sigma_i)$, $J^{\mu,\nu}$, E, E'. So, we get:

$$L_{1,2}(u_2) \in F_{\mu-\delta,\mu'}. \quad (3.42)$$

The equation (3.30) becomes, with (3.32), (3.41), (3.42),

$$L_{1,1}(w_1) = \frac{\partial f_1}{\partial u_1}(\phi_1, \phi_2)w_1 \ [F_{\mu-\delta,\mu'}]. \quad (3.43)$$

In local coordinates, where: $\Sigma_i = \{x_i = 0\}$, $(i = 1, 2)$, $L_{1,1} \equiv \partial_2 + b_1(x)$, with $b_1 \in C^\infty$, $x_0 = 0$, if we note $M_1 \equiv \frac{\partial f_1}{\partial u_1}(\phi_1, \phi_2) - b_1$, we find that w_1 satisfy:

$$(\partial_2 - M_1)w_1 = r_1 \in F_{\mu-\delta,\mu'}. \quad (3.44)$$

Now, we use the hypothesis of theorem 3.3.8 near C_1^-. Let $Z(x_2)$ the function defined in (3.22) and $y_1 = Z(x_2)w_1$. We get

$$(\partial_2 - M_1)y_1 = h_1 \in F_{\mu-\delta,\mu'},\ y_1 \equiv 0 \ \text{ for } \ x_2 < -\epsilon". \tag{3.45}$$

The integration of (3.45), using ($P_{3.5}$), gives $y_1 \in F_{\mu-\delta,\mu'}$. Then, $w_1 \in F_{\mu-\delta,\mu'}$ near x_0, and we have proved ($H_{\mu-\delta,\mu'}$) for u_1.

Second step: We prove ($H_{\mu-\delta,\mu'}$) for u_2.
Using (3.29), we find:

$$L_{2,2} \ = \ f_2(u) - L_{2,1}(u_2). \tag{3.46}$$

The same arguments developed for the linearization of $f_1(u)$, gives, using ($H_{\mu-\delta,\mu'}$) for u_1:

$$f_2(u) \ = \ \frac{\partial f_2}{\partial u_2}(\phi_1,\phi_2)w_2 \ \ [F_{\mu-\delta,\mu'}]. \tag{3.47}$$

We get also:

$$L_{2,1}(u_1) \ \in \ F_{\mu-\delta,\mu'}. \tag{3.48}$$

We find, with (3.46) – (3.48) :

$$L_{2,2}(u_2) \ = \ \frac{\partial f_2}{\partial u_2}(\phi_1,\phi_2)w_2 \ \ [F_{\mu-\delta,\mu'}],$$

so,

$$(L_{2,2} - \frac{\partial f_2}{\partial u_2}(\phi_1,\phi_2))u_2 \ \in \ F_{\mu-\delta,\mu'}. \tag{3.49}$$

In local coordinates, where: $\Sigma_i = \{x_i = 0\}$, $(i = 1,2)$, $L_{2,2} \equiv \partial_1 + b_2(x)$, with $b_2 \in C^\infty$, $x_0 = 0$, (3.49) becomes:

$$(\partial_1 - M_2)u_2 \ = \ r_2 \in \ F_{\mu-\delta,\mu'}, \tag{3.50}$$

with $M_2 \ \equiv \ \frac{\partial f_2}{\partial u_2}(\phi_1,\phi_2) - b_2$.

We conclude as in the first step, introducing the function $Z(x_1)$ and integrating the equation deduced from (3.50).

We find $u_2 \in F_{\mu-\delta-1,\mu'}$. We have proved $(H_{\mu-\delta,\mu'})$ for u_2, and so, by induction, we get: $\forall \mu$, $\forall \mu'$, $u \in F_{\mu,\mu'}$. Then $u \in P(E, E')$, which prove theorem 3.3.8.

Proof of theorem 3.3.7:

The first part is the particular case of theorem 3.2.7 for $E = I^{\mu}(\Sigma_1)$, $E' = I^{\mu}(\Sigma_2)$. The proof of the second part follows from the proof of theorem 3.3.8. Lemma 3.3.9, with $\mu = \mu_0$, $\mu' = \mu_0'$ gives the polarization of the solution. The proof of $(H_{\mu-\delta,\mu'})$ (respectively $(H_{\mu,\mu'-\delta})$) gives the transport equation in theorem 3.3.7 for $j = 1$ (respectively $j = 2$). Precisely, the transport equation for $j = 1$, for example, is an immediate consequence of (3.43) with $u_1 = \phi_1 + w_1$ $(v_1 \equiv w_{1,0} \equiv 0)$.

Chapter 4

Symbolic calculus: interaction for $N \times N$ systems.

H. BOUGRINI

4.1 Introduction

We saw in chapter 3 how the Bony's Theorem 2.1.2 could be precised at the symbolic level in the case of 2×2 systems. The main tool for this study wase the use of the conormal algebras $J^{\mu,\mu'}(\Sigma_1, \Sigma_2)$, which are completions of the tensor produit $I^{\mu}(\Sigma) \otimes I^{\mu'}(\Sigma')$.
In the general case of $N \times N$ systems involving more than 2 characteristic hypersurfaces Σ_j through Γ, we will use conormal algebras $I^{\mu,\nu}(\Sigma_j, \Gamma)$ in which the regularity at the wedge plays a crucial role: The index μ stills correspond to the regularity $I^{\mu}(\Sigma_j)$ outside Γ, and the new index ν precises an additionnal microlocal regularity on the conormal bundle to the wedge. For instance, the spaces $H_s(\Sigma_j)$ and $H_s(\Sigma_j, \Gamma)$ introduced at the chapter 2 correspond to $\mu = -s - \frac{1}{2}$ and respectively to $\nu = -\infty$ and ν close to $-\frac{1}{2}$ (see $(P_{4,2})$).
We assume here the geometrical hypothesis stated just before the Theorem 2.1.2, and we consider a priori a solution u of the system (2.1) $Lu = f(x, u)$ belonging to the conormal algebra $H_s(\Sigma_1 \cup .. \cup \Sigma_r)$ with $s > 1$ (see Remark 1.2.14). For technical simplification, we shall assume that the coefficients A_j of L are real matrices, and that the solution u is

$I\!R^N$ valued. Since the properties we are interested in are known outside the wedge Γ (thanks to the results of chapter 3), we will work close to a point $x_0 \in \Gamma$. Let C_j $(1 \leq j \leq r)$ be the ray on Σ_j through x_0, and denote C_j^+, C_j^- the two connected component of $C_j \setminus \{x_0\}$.
We assum

$$\begin{cases} u \in H_s(\Sigma_1) & \text{close to } C_1^- \\ u \in H_{s'}(\Sigma_2) & \text{close to } C_2^- \text{ with } s' \geq s \\ u \in C^\infty & \text{close to } C_j^- \text{ if } 3 \leq j \leq r. \end{cases}$$

We firstly study the regularity of u close to C_j^+ for $3 \leq j \leq r$: we prove that $u \in H_{s+s'+\frac{1}{2}}(\Sigma_j)$ close to these points - Note that the results of J. M. Bony $[Bo1, 2]$ and Wang Weike [W] gave only $u \in H_{s+s'-1}(\Sigma_j)$. Then we prove that our result is generically optimal, by giving a precise description of the outgoing principal symbols on the C_j^+, by means of the two incoming principal symbols on C_1^+ and C_2^+.
For $j = 1, 2$, the principal symbols on $C_j \setminus \{x_0\}$ have a linear behaviour like the ones in chapter 3: they satisfy across Γ the standard transport equation corresponding formally to the linearization of the system (2.1).
For $j \geq 3$, the principal symbols, which are 0 on C_j^-, still satisfy the standard transport equation on $C_j^+ \cup \{x_0\}$. Morover, they satisfy an intial condition at x_0 described in Theorem 4.4.3. This initial condition is in general different from 0, and allows us to link the effective non linearity of the source term of (2.1) to the appearance of singularities of "non linear type" on the C_j^+ for $3 \leq j \leq r$. Note that the results of J. M. Bony $[Bo1, 2]$ and Wang Weike $[W]$ did not guarantee that such "non linear" singularities really occur.
Moreover, our results allow a very accurate description of the principal part of these singularities, as shown in some examples at the last paragraph.

4.2 Conormal distributions precised at the wedge

The most part of the results stated in this paragraph are established in A. Piriou and J. P. Varenne $[PV2]$, Greenleaf A., Uhlmann G. $[GU]$, H. Bougrini , A. Piriou and J. P. Varenne $[BPV]$, G. Laschon $[La]$.
For μ, $\nu \in I\!R$, we denote by $S^{\mu,\nu}$ the space of all symbols $a(x, \xi_1, \xi_2)$ in $C^\infty(I\!R^n \times I\!R \times I\!R)$ such that :

$\forall \beta, j, k$, $\forall K$ compact in $\mathbb{R}^n$, $\exists C \geq 0$ with

$$|\partial_x^\beta \partial_{\xi_1}^j \partial_{\xi_2}^k a| \leq C(1 + |\xi_1| + |\xi_2|)^{\mu - j}(1 + |\xi_2|)^{\nu - k}. \tag{4.1}$$

We consider the distribution u in $\mathbb{R}^n$ defined by the oscillatory integral

$$u(x) = \int e^{i(x_1\xi_1 + x_2\xi_2)} a(x, \xi_1, \xi_2) d\xi_1 d\xi_2 \tag{4.2}$$

with $a \in S^{\mu,\nu}$.
Let $\Sigma = \{x_1 = 0\}$, $\Gamma = \{x_1 = x_2 = 0\}$. It is easy to check that u is C^∞ outside Σ, $u \in I^\mu(\Sigma)$ close to each point of $\Sigma \setminus \Gamma$ and that a property of invariance by diffeomorphisms preserving Σ, Γ allows to give the following definition

Definition 4.2.1 *Let Ω be open in $\mathbb{R}^n$, Σ a closed hypersurface in Ω, Γ a closed hypersurface in Σ. For $\mu, \nu \in \mathbb{R}$, we denote by $I^{\mu,\nu}(\Sigma, \Gamma)$ the space of all distributions u in Ω such that :*
u is C^∞ outside Σ, $u \in I^\mu(\Sigma)$ close to each point of $\Sigma \setminus \Gamma$, and, close to each point of Γ, for local coordinates such that $\Sigma = \{x_1 = 0\}$, $\Gamma = \{x_1 = x_2 = 0\}$, u is the form (4.2) with $a \in S^{\mu,\nu}$.

It can be proven that (4.2) can be rewritten with a symbol $a \in S^{\mu,\nu}$ independant of the variables x_1, x_2.

Now we give some linear properties of these spaces

($\mathbf{P_{4,1}}$) *If $v \in I^{\mu,\nu}(\Sigma, \Gamma)$, then $WF(v) \subset N(\Sigma) \cup N(\Gamma)$ and, microlocally in $N(\Gamma) \setminus N(\Sigma)$, $v \in I^{\mu+\nu}(\Gamma)$ (see the notation (1.32)).*

($\mathbf{P_{4,2}}$) *For $\varepsilon > 0$, we have $H_s(\Sigma, \Gamma) \subset I^{\mu, -\frac{1}{2}}(\Sigma, \Gamma) \subset H_{s-\varepsilon}(\Sigma, \Gamma)$ (see Definition (1.32)).*

($\mathbf{P_{4,3}}$) *If $v \in I^{\mu,\nu}(\Sigma, \Gamma)$ and $\mu < -1$, $\mu + \nu < -2$, then v is a continuous function, and we have the following trace properties:*
i) $v_{|\Sigma} \in I^{\mu+\nu+1}(\Gamma)$
ii) *If Σ' is a closed hypersurface in Ω meeting transversally Σ along Γ, then $v_{|\Sigma'}$ belongs to $I^\mu(\Gamma)$ if $\nu < -1$, to $I^{\mu+\varepsilon}(\Gamma)$ if $\nu = -1$, $\varepsilon > 0$, and to $I^{\mu+\nu+1}(\Gamma)$ if $\nu > -1$.*

Conversely, we can state an extension property:

($\mathbf{P}_{4,4}$) *If φ is a function in Σ belonging to $I^{\mu}(\Gamma)$ with $\mu < -1$, then there exists an extension ψ of φ in Ω such that $\psi \in I^{\mu-1}(\Gamma)$.*

From this we deduce

($\mathbf{P}_{4,5}$) *If $v \in I^{\mu,\nu}(\Sigma,\Gamma)$ with $\mu < -1$, $\mu+\nu < -2$, $\nu \leq 0$, then we can write $v = v_1 + v_2$ with $v_1 \in I^{\mu+\nu}(\Gamma)$ and $v_2 \in I^{\mu,\nu}(\Sigma,\Gamma)$, $v_2 = 0$ on Σ.*

The action of pseudodifferential operators on $I^{\mu,\nu}(\Sigma,\Gamma)$ is described as follows:

($\mathbf{P}_{4,6}$) *If $P \in \Psi^{d}(\Omega)$ and $u \in I^{\mu,\nu}(\Sigma,\Gamma)$, then $Pu \in I^{\mu+d,\nu}(\Sigma,\Gamma)$. If moreover $P \in \Psi^{-\infty}$ in a conic neighbourhood of $N(\Sigma)$, then $Pu \in I^{\mu+\nu+d}(\Gamma)$.*

In particular, if V is a vector field in Ω and $u \in I^{\mu,\nu}(\Sigma,\Gamma)$, we get

$$(\mathbf{P}_{4,7}) \begin{cases} Vu \in I^{\mu+1,\nu}(\Sigma,\Gamma) & \\ Vu \in I^{\mu,\nu+1}(\Sigma,\Gamma) & \text{if } V \text{ is tangent to } \Sigma \\ Vu \in I^{\mu+1,\nu-1}(\Sigma,\Gamma) & \text{if } V \text{ is tangent to } \Gamma \\ Vu \in I^{\mu,\nu}(\Sigma,\Gamma) & \text{if } V \text{ is tangent to } \Sigma \text{ and } \Gamma. \end{cases}$$

We will often use the following properties, when $u \in I^{\mu,\nu}(\Sigma,\Gamma)$, $\mu < -1$, $\mu+\nu < -2$, $\nu \leq 0$.

($\mathbf{P}_{4,8}$) *If $u = 0$ on Σ, then, locally for $\Sigma = \{x_1 = 0\}$, one can write $u = x_1 v$ with $v \in I^{\mu+1,\nu}(\Sigma,\Gamma)$.*

($\mathbf{P}_{4,9}$) *If $u = 0$ on Γ, then, locally for $\Gamma = \{x_1 = x_2 = 0\}$, one can write $u = x_1 u_1 + x_2 u_2$ with $u_1, u_2 \in I^{\mu+1,\nu}(\Sigma,\Gamma)$.*

($\mathbf{P}_{4,10}$) *Locally for $\Sigma = \{x_1 = 0\}$, $\Gamma = \{x_1 = x_2 = 0\}$, one has $x_1 u \in I^{\mu-1,\nu}(\Sigma,\Gamma)$, $x_2 u \in I^{\mu,\nu-1}(\Sigma,\Gamma)$.*

($\mathbf{P}_{4,11}$) *Let $\Sigma = \{x_1 = 0\}$, $\Gamma = \{x_1 = x_2 = 0\}$, and let Σ' be a closed hypersurface meeting transversally Σ along Γ. Let $v \in I^{\mu,\nu}(\Sigma,\Gamma)$ or $v \in I^{\mu,\nu}(\Sigma',\Gamma)$ with $\mu < -1$, $\mu+\nu < -2$. We set $V(x) = \int_{\alpha}^{x_2} v(x_1, s, x'')ds$, where $\alpha < 0$ and x_2 are sufficiently close to 0.*
Then

a) If $v \in I^{\mu,\nu}(\Sigma',\Gamma)$, we have $V \in I^{\mu-1,\nu}(\Sigma',\Gamma) + I^{\mu+\nu,-1}(\Sigma,\Gamma)$. Moreover, if v is given by (4.2) with $(x'',\xi_1,\xi_2) \longmapsto a(x'',\xi_2,\xi_1)$ is in $S^{\mu,\nu}$ and independant of x_1, x_2, close to each point of $\Sigma^+ = \{x_1 = 0, x_2 > 0\}$ we have

$$V(x) \equiv \int e^{ix_1\xi_1} a(x'',\xi_1,0)d\xi_1 \quad mod. \ \ C^\infty$$

b)If $v \in I^{\mu,\nu}(\Sigma',\Gamma)$ and $WF(v) \cap N(\Sigma) = \emptyset$, we have $V \in I^{\mu-1,\nu}(\Sigma',\Gamma)$.
c)If $v \in I^{\mu,\nu}(\Sigma,\Gamma)$, we have $V \in I^{\mu,\nu-1}(\Sigma,\Gamma)$.

The property $(P_{4,11}; a)$ is crucial for our symbolic study, and we give a sketch of the proof.
We can suppose $\Sigma' = \{x_2 = 0\}$, $a(x'',\xi_1,\xi_2) = 0$ if $|\xi_1| + |\xi_2| \leq 1$. Let $\chi \in C^\infty(I\!R^2 \setminus 0)$ be homogeneous of degree 0, with $\chi = 1$ if $|\xi_2| \leq \varepsilon_1|\xi_1|$, $\chi = 0$ if $\xi_2| \geq \varepsilon_2|\xi_1|$, $0 < \varepsilon_1 < \varepsilon_2$. Then $V = V_1 + V_2$ where

$$V_1(x) = \int_\alpha^{x_2} \int\int e^{i(x_1\xi_1 + x_2\xi_2)} \chi(\xi_1,\xi_2) a(x'',\xi_1,\xi_2) d\xi_1 d\xi_2 ds \tag{4.3}$$

and where $V_2(x)$ is defined in the same way, just replacing χ by $1 - \chi$. Remarking that $|\xi_2| \geq \varepsilon_1|\xi_1|$ on the support of $1-\chi$ and integrating with respect to s, we obtain $V_2 = W - W_0$ where

$$W(x) = \int\int e^{i(x_1\xi_1 + x_2\xi_2)} a_2(x'',\xi_1,\xi_2) d\xi_1 d\xi_2,$$

$a_2(x'',\xi_1,\xi_2) = a(x'',\xi_1,\xi_2)\frac{1-\chi(\xi_1,\xi_2)}{i\xi_2}$, $W_0(x) = W(x_1,\alpha,x'')$.
We have $W \in I^{\mu-1,\nu}(\Sigma',\Gamma)$, $W_0 \in C^\infty$, thus $V_2 \in I^{\mu-1,\nu}(\Sigma',\Gamma)$.
For V_1, we remark that χa is an ordinary symbol of degree $\mu + \nu$: $\chi a \in S^{\mu+\nu}(I\!R^{n-1} \times I\!R^2)$. Let $\psi \in C_0^\infty(I\!R)$ such that $\psi = 1$ close to 0. The decomposition $\chi a = a' + a''$ with $a' = \chi a \psi(\xi_2)$, $a'' = \chi a(1-\psi)(\xi_2)$ gives rise to a decomposition $V_1 = V_1' + V_2''$. Since $a' \in S^{\mu+\nu,-\infty}$, its contribution in v belongs to $I^{\mu+\nu}(\Sigma)$ and can be rewritten in the form $v_1'(x) = \int e^{ix_1\xi_1} b(x',\xi_1)d\xi_1$ where $b \in S^{\mu+\nu}$, $x' = (x_2,x'')$. By integrating b with respect to x_2, we obtain $V_1' \in I^{\mu+\nu}(\Sigma)$. For the term V_1'', an integration with respect to s gives easily $V_1'' \in I^{\mu+\nu,-1}(\Sigma,\Gamma)$, and we have proved the first part of $(P_{4,11}; a)$, since it is not hard to prove that any distribution of the form (4.2) can be locally rewritten with a symbol independant of x_1, x_2. For the second part of $(P_{4,11}; a)$, Taylor's formula gives

$$\chi(\xi_1,\xi_2)a(x'',\xi_1,\xi_2) = \chi(\xi_1,\xi_2)a(x'',\xi_1,0) + \xi_2\chi(\xi_1,\xi_2)r(x'',\xi_1,\xi_2)$$

where $\chi(\xi_1,\xi_2)r(x'',\xi_1,\xi_2) \in S^{\mu+\nu-1}(\mathbb{R}^{n-2}\times\mathbb{R}^2)$, and the term $\xi_2\chi r$ gives rise to a contribution in (4.3) belonging to $I^{\mu+\nu-1}(\Gamma)$, which is C^∞ outside Γ. For the contribution of the other term in (4.3), we write

$$\chi(\xi_1,\xi_2)a(x'',\xi_1,0) = a(x'',\xi_1,0) + (\chi(\xi_1,\xi_2)-1)a(x'',\xi_1,0)$$

The last term gives in (4.3) a contribution belonging to $I^{-1,\mu+\nu}(\Sigma',\Gamma)$, which is C^∞ close to Σ^+. The first term gives the explicit contribution

$$\int e^{ix_1\xi_1}\Big(\int (e^{ix_2\xi_2}-e^{ia\xi_2})\frac{d\xi_2}{i\xi_2}\Big)a(x'',\xi_1,0)d\xi_1 =$$

$$2\pi\mathcal{H}(x_2)\int e^{ix_1\xi_1}a(x'',\xi_1,0)d\xi_1,$$

where $\mathcal{H}$ is the Heaviside function. This completes the proof of $(P_{4,11};a)$.

In the situation (1.8) where r hypersurfaces $\Sigma_1,..,\Sigma_r$ meet transversally along Γ, we set

$$I^{\mu,\nu}(\Sigma_1,..,\Sigma_r) = \sum_{j=1}^{r} I^{\mu,\nu}(\Sigma_i,\Gamma).$$

Note that the space $J^{\mu,\mu'}(\Sigma_1,\Sigma_2)$ introduced in Chapter 3 (see Definition 3.3.1) coincides with $I^{\mu,\mu'}(\Sigma_1,\Gamma)+I^{\mu',\mu}(\Sigma_2,\Gamma)$ if $\mu,\ \mu'\leq 0$.

We conclude this paragraph with some nonlinear properties of our spaces of conormal distributions

($\mathbf{P_{4,12}}$) *The stability under C^∞ functions (like in Proposition 1.2.6) is satisfied in the following cases*

$$\mu<-1,\ \mu+\nu<-2,\ \nu<-\frac{1}{2}\ \textit{for}\ I^{\mu,\nu}(\Sigma,\Gamma)$$

$$\mu<-1,\ \mu+\nu<-2,\ \mu\leq\nu<-\frac{1}{2}\ \textit{for}\ I^{\mu,\nu}(\Sigma_1,..,\Sigma_r).$$

($\mathbf{P_{4,13}}$) *The stability by multiplication of the space $\sum_{i=1}^{r} I^{\mu_i,\nu_i}(\Sigma_i,\Gamma)$ is true if $\mu_i<-1$, $\mu_i+\nu_i<-2$, $\nu_i\leq 0$ for all i and if $Max(\mu_i,\mu_i+\nu_i+1)\leq \nu_j$ for all $j\neq i$ such that $\nu_i\leq 0$, $\nu_i\neq -1$, $\mu_i<\nu_j$ for all $j\neq i$ such that $\nu_i=-1$.*

($\mathbf{P_{4,14}}$) *The stability of $\sum_{i=1}^{r} I^{\mu_i,\nu_i}(\Sigma_i,\Gamma)$ under C^∞ functions is true* *if we add the condition $\nu_j < -\frac{1}{2}$ for all j.*

Multiplication may have a regularizing effect:

($\mathbf{P_{4,15}}$) *If $\mu, \nu < -1$ and if $u, v \in I^{\mu,\nu}(\Sigma, \Gamma)$ vanish on Σ, then $uv \in I^{\mu-\delta,\nu}(\Sigma, \Gamma)$, $\delta = Min(1, -(\mu+1))$ if $\mu \neq -2$, δ arbitrary in $]0, 1[$ if $\mu = -2$.*

($\mathbf{P_{4,16}}$) *If $\mu < -1$ and if $u, v \in I^{\mu,-1}(\Sigma, \Gamma)$ vanish on Σ, then $uv \in I^{\mu-\delta,-1+\varepsilon}(\Sigma, \Gamma)$ where $\varepsilon > 0$ and δ defined in $(P_{4,15})$.*

Finally we state a result of unicity related to the definition of $I^{\mu,\nu}(\Sigma_1, .., \Sigma_r)$

($\mathbf{P_{4,17}}$) *If $\sum_{i=1}^{r} w_i = \sum_{i=1}^{r} w_i^{'}$ with $w_i \in I^{\mu_i,\nu_i}(\Sigma_i, \Gamma)$ and $w_i^{'} \in I^{\mu_i^{'},\nu_i^{'}}(\Sigma_i, \Gamma)$, then $w_i \equiv w_i^{'}$ modulo $I^{\rho}(\Gamma)$ for $i = 1, .., r$ where $\rho = Max_i(\mu_i + \nu_i, \mu_i^{'} + \nu_i^{'})$.*

4.3 Conormal regularity of the solution

In the context of the Theorem of J.M. Bony 2.1.2, we consider a $\mathbb{R}^N$ valued solution u of the semilinear system

$$Lu = f(x, u) \quad (\text{ see } (2.1)) \tag{4.4}$$

belonging to the conormal space $H_s(\Sigma_1, .., \Sigma_r)$. We shall assume that $s > 1$ and that the conormal spaces are based on the Besov spaces. We shall prove that the regularity of the solution is in fact stronger than the regularity described by the mere space $H_s(\Sigma_1, .., \Sigma_r)$: we have an additional regularity at the wedge, combining with an additional regularity on Σ_1,..,Σ_r outside the wedge which is expected from the results of Chapter 3 about the case of a single hypersurface, and particulary from the polarization property of the principal symbol (see Theorem 3.2.1). As already pointed out in introduction, we work close to a point of the wedge Γ. To take in account simultaneously the polarization properties associated to Σ_1,..,Σ_r, we consider the Reduction 2.3.2 of L. In order to avoid too much technical considerations, we shall assume that $r = N$, that is:

The number r of characteristic hypersurfaces issued from the wedge Γ is equal to the size N of the system L.

We refer to Laschon G. [*La*] for the adaptation of our method to the case $r < N$. This adaptation is sketched in the Introduction of Chapter 5.

Considering Reduction 2.3.2 in the case $r = N$, recall that $(e_1, .., e_N)$ (resp.$(e'_1, .., e'_N)$) is a basis of $\mathbb{R}^N$, depending smoothly on x, such that $e_j(x)$ (resp.$e'_j(x)$) is a basis of $\text{Ker}l(x,\xi)$ (resp. $\text{Ker}^t l(x,\xi)$) when $(x,\xi) \in N(\Sigma_j)$. We have denoted by $l(x,\xi) = \sum_{j=1}^n \xi_j A_j(x)$ the principal symbol of L.
We set $u = \sum_{j=1}^N u_j e_j = (u_1, .., u_N)$.
The system (4.4) is then equivalent to

$$\sum_{k=1}^{N} L_{j,k} u_k = f_j(x,u) \qquad (4.4)_j$$

for $j = 1, .., N$, where $f_j =< e'_j, f >$, $L_{j,k} u_k =< e'_j, L(u_k e_k >$.

From Proposition 2.3.3 we know that the principal part of $L_{j,k}$ is tangent to Σ_j, Σ_k, Γ when $j \neq k$, and, in the case $j = k$: tangent to Σ_j, transverse to Γ, and proportional to the projection on $T_x(\Omega)$ of the hamiltonian field $H_p(x,\xi)$ when $(x,\xi) \in N(\Sigma_j)$.

We have also

$(\mathbf{P_{4,18}})$ *For $i = 1, .., N$, the system $L'_i = (L_{j,k})_{j,k \neq i}$ is elliptic on $N(\Sigma_i)$*

The following result gathers the polarization properties associated to $\Sigma_j \setminus \Gamma$ and the regularity at the wedge Γ.

Proposition 4.3.1 *Let u be a solution of $(4.4)_j$ ($1 \leq j \leq N$), belonging to the conormal Besov based space $H_s(\Sigma_1, .., \Sigma_N)$ with $s > 1$. Then $u \in I^{\mu,\mu-1}(\Sigma_1, .., \Sigma_N)$ if $\mu = -s - \frac{1}{2}$, and for $j = 1, .., N$:*

$$u_j \in I^{\mu,\mu-1}(\Sigma_j, \Gamma) + \sum_{k \neq j} I^{\mu-1,\mu}(\Sigma_k, \Gamma).$$

Proof. First step: regularity at the wedge
Assume that we have the induction hypothesis

$$u,\ f(x,u) \in I^{\mu,\nu}(\Sigma_1, .., \Sigma_N)\ ,\ \nu \leq 0,\ \mu + \nu < -2 \qquad (4.5)$$

Note that this hypothesis is satisfied for $\nu = -\frac{1}{2}$ thanks to $(P_{4,2})$ and Remark 1.2.14.

We shall prove that that

$$u \in I^{\mu,\nu-1}(\Sigma_1, .., \Sigma_N) \tag{4.6}$$

Let $j \in \{1, .., N\}$. A change of local coordinates allows to write $\Sigma_j = \{x_1 = 0\}$, $\Gamma = \{x_1 = x_2 = 0\}$, $L_{j,j} = \partial_{x_2} + b$ with $b \in C^\infty$, and thus the equation $(4.4)_j$ takes the form

$$\partial_{x_2} u_j = g_j \ , \ \ g_j = f_j(x, u) - bu_j - \sum_{k \neq j} L_{j,k} u_k.$$

Property $(P_{4,7})$ show that

$$g_j \in I^{\mu,\nu}(\Sigma_j, \Gamma) + \sum_{k \neq j} I^{\mu+1,\nu-1}(\Sigma_k, \Gamma).$$

If $\alpha < 0$ is sufficiently close to 0, we have

$$u_j(x) = u_j(x_1, \alpha, x") + \int_\alpha^{x_2} g_j(x_1, s, x")ds. \tag{4.7}$$

Since the function $x \longmapsto u_j(x_1, \alpha, x")$ is obviously in $I^\mu(\Sigma_j)$ $(P_{4,11})$ yields

$$u_j \in I^{\mu,\nu-1}(\Sigma_j, \Gamma) + \sum_{k \neq j} I^{\mu,\nu-1}(\Sigma_k, \Gamma) + I^{\mu+\nu,-1}(\Sigma_j, \Gamma).$$

But $I^{\mu+\nu,-1}(\Sigma_j, \Gamma) \subset I^{\mu,\nu-1}(\Sigma_j, \Gamma)$,and (4.6) follows. Using $(P_{4,12})$, an induction gives finally

$$u \in I^{\mu,\mu-1}(\Sigma_1, .., \Sigma_N) \tag{4.8}$$

Second step: Property $(P_{4,18})$ allows to consider a microlocal parametrix $P_i \in \Psi^{-1}$ of $L_i^{'}$, with $P_i L_i^{'} = I + R_i$, $R_i \in \Psi^0$ and of order $-\infty$ in a conical neighbourhood of $N(\Sigma_i)$.
Setting $w_i = (u_k)_{k \neq i}$, (4.4) gives $L_i^{'} w_i = h$, $h_k = f_k(x, u) - L_{k,i} u_i$ for $1 \leq k \leq N$ and $k \leq i$.
Properties $(P_{4,7})$, $(P_{4,12})$ gives

$$h_k \in I^{\mu,\mu}(\Sigma_i, \Gamma) + \sum_{j \neq i} I^{\mu+1,\mu-1}(\Sigma_j, \Gamma).$$

Since $w_i = P_i h - R_i w_i$, $(P_{4,6})$ yields

$$w_i \in I^{\mu-1,\mu}(\Sigma_i, \Gamma) + \sum_{j \neq i} I^{\mu,\mu-1}(\Sigma_j, \Gamma) \tag{4.9}$$

and the last assertion of Proposition 4.3.1 follows from $(P_{4,5})$ and $(P_{4,17})$.

Now consider a point $x_0 \in \Gamma$ and let C_j $(1 \leq j \leq N)$ be the ray on Σ_j through x_0. We give a refinement of Proposition 4.3.1. We denote by C_j^- , C_j^+ the two connected components of $C_j \setminus \{x_0\}$.

Proposition 4.3.2 *Let u be a solution of $(4.4)_j$ $(1 \leq j \leq N)$, belonging to the conormal Besov based space $H_s(\Sigma_1, .., \Sigma_N)$ with $s > 1$. Assume that $u \in I^{\mu'}(\Sigma_j)$ close to the points of C_j^- $(2 \leq j \leq N)$ with $\mu' \leq \mu$, $\mu = -s - \frac{1}{2}$.*
Then, close to x_0:

$$u_1 \in I^{\mu,\mu'-1}(\Sigma_1, \Gamma) + I^{\mu'-1,\mu}(\Sigma_2, .., \Sigma_N)$$

and, for $j \geq 2$:

$$u_j \in I^{\mu',\mu-1}(\Sigma_j, \Gamma) + I^{\mu-1,\mu'}(\Sigma_1, \Gamma) + I^{\mu'-1,\mu}(\Sigma'_{1,j})$$

where $I^{\mu'-1,\mu}(\Sigma'_{1,j}) = \sum_{k \neq 1,j} I^{\mu'-1,\mu}(\Sigma_k, \Gamma)$.

Proof.
Let us consider the property

$$(\mathbf{R}_\lambda) \begin{cases} u_1 \in I^{\mu,\lambda-1}(\Sigma_1, \Gamma) + I^{\lambda-1,\mu}(\Sigma_2, .., \Sigma_N) \\ u_j \in I^{\lambda,\mu-1}(\Sigma_j, \Gamma) + I^{\mu-1,\lambda}(\Sigma_1, \Gamma) + I^{\lambda-1,\mu}(\Sigma'_{1,j}) \quad \text{if } j = 2, .., N \\ \text{with } \mu' \leq \lambda \leq \mu \end{cases}$$

Proposition 4.3.1 gives (R_μ), and our goal is to establish $(R_{\mu'})$. We define θ as follows

$$\theta = Min(1, -(\lambda + 1)) \text{ if } \lambda \neq -2, \ \ \theta \in]0, 1[\text{ if } \lambda = -2 \qquad (4.10)$$

We shall prove that (R_λ) implies $(R_{\lambda-\theta})$ if $\lambda - \theta \geq \mu'$, which obviously gives $(R_{\mu'})$ by induction. Assume that (R_λ) is satisfied

First step: Improvment of the Σ_j-index of u_j if $j \geq 2$.
We prove that

$$u_j \in I^{\lambda-\theta,\mu}(\Sigma_j, \Gamma) + I^{\mu-1,\lambda}(\Sigma_1, \Gamma) + I^{\lambda-1,\mu}(\Sigma'_{1,j}) \quad \text{if } j \geq 2. \quad (4.11)$$

The strategy consists in a linearization of the equation $(4.4)_j$ giving an ordinary differential equation for u_j with good coefficients, and to use the

regularity $I^{\mu'}(\Sigma_j)$ in the past.
We can write

$$u_j = \psi_j + v_j\ ,\quad \psi_j \in I^{\mu-1,\lambda}(\Sigma_1,\Gamma) + I^{\lambda-1,\mu}(\Sigma'_{1,j}) \qquad (4.12)$$
$$v_j \in I^{\lambda,\mu-1}(\Sigma_j,\Gamma)$$

From $(P_{4,3})$ we know that the trace of v_j on Σ_j is in $I^{\lambda+\mu}(\Gamma)$ and it can be extended (see $(P_{4,4})$ to an element of $I^{\lambda+\mu-1}(\Gamma)$ defined in a neighbourhood of x_0 in Ω. Since $I^{\lambda+\mu-1}(\Gamma) \subset I^{\mu-1,\lambda}(\Sigma_1,\Gamma)$, we can modify ψ_j in such a way that (4.12) is satisfied and $v_j = 0$ on Σ_j.
Using this decomposition of u_j and denoting $\gamma_j = (u_1, .., \psi_j, .., u_N)$, Taylor's formula yields $f_j(x,u) = \frac{\partial f_j}{\partial u_j}(x,\gamma)u_j + R_j$, $R_j = f_j(x,\gamma_j) - \frac{\partial f_j}{\partial u_j}(x,\gamma_j)\psi_j + v_j^2 R(x,u,\psi_j)$, where R_j is a C^∞ function.
Thanks to $(P_{4,12})$, $(P_{4,13})$, $(P_{4,14})$ we get $f_j(x,\gamma_j) - \frac{\partial f_j}{\partial u_j}(x,\gamma_j)\psi_j \in I^{\lambda-1,\mu}(\Sigma_j,\Gamma) + I^{\mu,\lambda}(\Sigma_1,\Gamma) + I^{\lambda,\mu}(\Sigma'_{1,j})$.
Similary u and ψ_j are in $I^{\mu,\lambda-1}(\Sigma_1,\Gamma) + I^{\lambda,\mu-1}(\Sigma_2, .., \Sigma_N)$, thus one can write $R(x,u,\psi_j) = A + B$ with $A \in I^{\mu,\lambda}(\Sigma_1,\Gamma) + I^{\lambda,\mu}(\Sigma'_{1,j})$, $B \in I^{\lambda,\mu}(\Sigma_j,\Gamma)$. Since $v_j \in I^{\lambda,\mu-1}(\Sigma_j,\Gamma)$ vanishes on Σ_j, $(P_{4,15})$ gives $v_j^2 \in I^{\lambda-\theta,\mu-1}$ where θ is defined by (4.10), therefore $v_j^2 A \in I^{\lambda-\theta,\mu}(\Sigma_j,\Gamma) + I^{\mu,\lambda}(\Sigma_1,\Gamma) + I^{\lambda,\mu}(\Sigma'_{1,j})$.
In the same way $(P_{4,15})$ shows that $v_j^2 B = v_j(v_j B) \in I^{\lambda-\theta,\mu}(\Sigma_j,\Gamma)$.

Using the equation $(4.4)_j$ with local coordinates such that $\Sigma_j = \{x_1 = 0\}$, $\Gamma = \{x_1 = x_2 = 0\}$, $L_{j,j} = \partial_{x_2} + b$ $(b \in C^\infty)$, $C_j^- \subset \{x_2 < 0\}$, and remarking that

$$\sum_{k \neq j} L_{j,k}u_k \in I^{\lambda-1,\mu}(\Sigma_j,\Gamma) + I^{\mu,\lambda-1}(\Sigma_1,\Gamma) + I^{\lambda,\mu-1}(\Sigma'_{1,j})$$

thanks to $(P_{4,7})$, we obtain the following linearization

$$(\partial_{x_2} - \varphi_j)u_j \;=\; F_j\ , \qquad (4.13)$$

$\varphi_j = \frac{\partial f_j}{\partial u_j}(x,\gamma_j) - b \in I^{\lambda-1,\mu}(\Sigma_j,\Gamma) + I^{\mu,\lambda}(\Sigma_1,\Gamma) + I^{\lambda,\mu}(\Sigma'_{1,j})$
$F_j = R_j - \sum_{k\neq j} L_{j,k}u_k \in I^{\lambda-\theta,\mu}(\Sigma_j,\Gamma) + I^{\mu,\lambda}(\Sigma_1,\Gamma) + I^{\lambda,\mu}(\Sigma'_{1,j})$
If we choose $\alpha < 0$ sufficiently close to 0 and if we set $G_j(x) = \int_\alpha^{x_2} \varphi_j(x_1,s,x'')ds$, explicit integration of (4.13) yields

$$u_j(x) \;=\; e^{G_j(x)} \int_\alpha^{x_2} (e^{-G}F_j)(x_1,s,x'')ds \;+\; e^{G_j(x)}u_j(x_1,\alpha,x'') \quad (4.14)$$

which proves (4.11) thanks to $(P_{4,11})$, since the function $x \longmapsto u_j(x_1, \alpha, x'')$ belongs to $I^{\mu'}(\Sigma_j)$ and $\mu' \leq \lambda - \theta$.

Second step: Improvement of the regularity of u at the wedge.
The method is the same as in the first step of the proof of Proposition 4.3.1. We have now:

$$f(x,u) \in I^{\mu,\lambda-\theta}(\Sigma_1,\Gamma) + I^{\lambda-\theta,\mu}(\Sigma_2,..,\Sigma_N) \tag{4.15}$$

$\sum_{k\neq 1} L_{1,k}u_k \in I^{\mu-1,\lambda}(\Sigma_1,\Gamma) + I^{\lambda,\mu-\theta}(\Sigma_2,..,\Sigma_N)$
and, if $j \neq 1$:
$\sum_{k\neq j} L_{j,k}u_k \in I^{\lambda-1,\mu}(\Sigma_j,\Gamma) + I^{\mu,\lambda-1}(\Sigma_1,\Gamma) + I^{\lambda,\mu-\theta}(\Sigma'_{1,j})$
thus (4.7), $(P_{4,11})$ give

$$u_1 \in I^{\mu,\lambda-1-\theta}(\Sigma_1,\Gamma) + I^{\lambda-1,\mu-\theta}(\Sigma_2,..,\Sigma_N) \tag{4.16}$$

and, if $j \neq 1$

$$u_j \in I^{\lambda-\theta,\mu-1}(\Sigma_j,\Gamma) + I^{\mu-1,\lambda-\theta}(\Sigma_1,\Gamma) + I^{\lambda-1,\mu-\theta}(\Sigma'_{1,j}). \tag{4.17}$$

Third step: improvement of the Σ_k-index of u_j when $k \neq 1, j$ (in the corresponding terms we want to transfer the gain θ from the second to the first index).
The method is the same as in the second step of the proof of Proposition 4.3.1. Using the notations introduced there, let P_k be a microlocal parametrix of L'_k in a conic neighbourhood of $N(\Sigma_k)$, $k \neq 1$ being fixed.
We have $L'_k w_k = h$, $h_j = f_j(x,u) - L_{j,k}u_k$ for $j \neq k$, thus $h \in I^{\mu,\lambda-\theta}(\Sigma_1,\Gamma) + I^{\lambda-\theta,\mu}(\Sigma_k,\Gamma) + I^{\lambda,\mu-\theta}(\Sigma'_{1,k})$ thanks to (4.15). Since (4.16), (4.17) show that $w_k \in I^{\mu,\lambda-\theta-1}(\Sigma_1,\Gamma) + I^{\lambda-1,\mu-\theta}(\Sigma_k,\Gamma) + I^{\lambda-\theta,\mu-1}(\Sigma'_{1,k})$, the equality $w_k = P_k h - R_k w_k$ yields now

$$u_j \in I^{\mu,\lambda-\theta-1}(\Sigma_1,\Gamma) + I^{\lambda-\theta-1,\mu}(\Sigma_k,\Gamma) + I^{\lambda-\theta,\mu-1}(\Sigma'_{1,k})$$

for $1 \leq j,k \leq N$, $k \neq 1$, $j \neq k$. Combining this with (4.16), (4.17) and using $(P_{4,17})$, we obtain $(R_{\lambda-\theta})$.

Now we give our regularity property in the case of interaction of two incoming conormal waves.

Theorem 4.3.3 *Let $x_0 \in \Gamma$, C_j the ray on Σ_j $(1 \leq j \leq N)$ through x_0, $C_j^{\pm}$ the two connected componants of $C_j \setminus \{x_0\}$. Let u be a $\mathbb{R}^N$-valued solution of $(4.4)_j$ $(1 \leq j \leq N)$, belonging to the conormal Besov based*

space $H_s(\Sigma_1,..,\Sigma_N)$ with $s > 1$.
Assume that $u \in I^{\mu'}(\Sigma_2)$ close to the point of C_2^- $(\mu' \leq \mu = -s - \frac{1}{2})$ and that $u \in C^\infty$ close to the point of C_j^- if $3 \leq j \leq N$.
Then, close to x_0:

$$u_1 \in I^{\mu,\mu'-1}(\Sigma_1,\Gamma) + I^{\mu'-1,\mu}(\Sigma_2,\Gamma) + I^{\mu+\mu'-1,-1}(\Sigma'_{1,2})$$

$$u_2 \in I^{\mu-1,\mu'}(\Sigma_1,\Gamma) + I^{\mu',\mu-1}(\Sigma_2,\Gamma) + I^{\mu+\mu'-1,-1}(\Sigma'_{1,2})$$

and, if $j \geq 3$:

$$u_j \in I^{\mu-1,\mu'}(\Sigma_1,\Gamma) + I^{\mu'-1,\mu}(\Sigma_2,\Gamma) + I^{\mu+\mu',-1}(\Sigma_j,\Gamma) + I^{\mu+\mu'-1,-1}(\Sigma'_{1,2,j})$$

where $I^{\alpha,\beta}(\Sigma'_{1,2}) = \sum_{k\neq 1,2} I^{\alpha,\beta}(\Sigma_k,\Gamma)$ and $I^{\alpha,\beta}(\Sigma'_{1,2,j}) = \sum_{k\neq 1,2,j} I^{\alpha,\beta}(\Sigma_k,\Gamma)$

Proof. We consider the property

$$(\mathbf{R}_{\lambda,\nu}) \begin{cases} u_1 \in I^{\mu,\mu'-1}(\Sigma_1,\Gamma) + I^{\mu'-1,\mu}(\Sigma_2,\Gamma) + I^{\lambda-1,\nu}(\Sigma'_{1,2}) \\ u_2 \in I^{\mu-1,\mu'}(\Sigma_1,\Gamma) + I^{\mu',\mu-1}(\Sigma_2,\Gamma) + I^{\lambda-1,\nu}(\Sigma'_{1,2}) \\ \text{and, if } j \geq 3 \\ u_j \in I^{\mu-1,\mu'}(\Sigma_1,\Gamma) + I^{\mu'-1,\mu}(\Sigma_2,\Gamma) + I^{\lambda,\nu}(\Sigma_j,\Gamma) + I^{\lambda-1,\nu}(\Sigma'_{1,2,j}) \end{cases}$$

with the conditions

$$\lambda \leq \mu' \leq \mu \leq \nu < -1 \ , \ \lambda + \nu \leq \mu + \mu' \tag{4.18}$$

Proposition 4.3.2 shows that $(R_{\lambda,\nu})$ and (4.18) are satisfied if $\lambda = \mu'$, $\nu = \mu$. Our goal is to establish $(R_{\mu+\mu',-1})$. We shall prove that $(R_{\lambda,\nu})$ implies $(R_{\lambda-\theta,\nu"})$ if $\lambda - \theta > \mu + \mu'$, θ defined by (4.10), $\nu" = Max(\nu,\nu')$, ν' defined by $\mu + \mu' - 1 = \lambda - \theta + \nu'$. The method is the same as in the proof of Proposition 4.3.2, thus we only give the sketch of the steps.

First step: Improvement of the Σ_j-index of u_j if $j \geq 3$.
We prove that, if $j \geq 3$:

$$u_j \in I^{\mu-1,\mu'}(\Sigma_1,\Gamma) + I^{\mu'-1,\mu}(\Sigma_2,\Gamma) + I^{\lambda-\theta,\nu"}(\Sigma_j,\Gamma) + I^{\lambda-1,\nu"}\Sigma'_{1,2,j} \tag{4.19}$$

As in the first step of the proof of Proposition 4.3.2, the argument is a linearization of the equation $(4.4)_j$. Remark also that (4.18) insures $I^{\lambda+\nu,-1}(\Sigma_j,\Gamma) \subset I^{\mu+\mu',-1}(\Sigma_j,\Gamma)$ and that $I^{\mu+\mu',-1}(\Sigma_j,\Gamma) \subset I^{\lambda-\theta,\nu'}(\Sigma_j,\Gamma)$ $\subset I^{\lambda-\theta,\nu"}(\Sigma_j,\Gamma)$ since $\lambda - \theta > \mu + \mu'$ and by definition of ν', $\nu"$.

Second step: Improvement of the regularity of u ay the wedge. Like in the first step of the proof of Proposition 4.3.1, using $(R_{\lambda,\nu})$, (4.19) and the inclusion $I^{\lambda-\theta+\nu",-1}(\Sigma_1,\Gamma) \subset I^{\mu,\mu'-1}(\Sigma_1,\Gamma)$, we get

$$u_1 \in I^{\mu,\mu'-1}(\Sigma_1,\Gamma) + I^{\mu'-1,\mu}(\Sigma_2,\Gamma) + I^{\lambda-1,\nu"-\theta}(\Sigma'_{1,2})$$

$$u_2 \in I^{\mu-1,\mu'}(\Sigma_1,\Gamma) + I^{\mu',\mu-1}(\Sigma_2,\Gamma) + I^{\lambda-1,\nu"-\theta}(\Sigma'_{1,2})$$

and if $j \geq 3$,

$$u_j \in I^{\mu-1,\mu'}(\Sigma_1,\Gamma) + I^{\mu'-1,\mu}(\Sigma_2,\Gamma) + I^{\lambda-\theta,\nu"}(\Sigma_j,\Gamma) + I^{\lambda-1,\nu"-\theta}(\Sigma'_{1,2,j}).$$

Third step: Improvment of the Σ_k-index of u_j when $k \neq 1,2,j$ (transfer of the gain θ). We use again an argument of microlocal ellipticity on $N(\Sigma_k)$.

We have thus obtained $(R_{\lambda-\theta,\nu"})$, and we remark that the new index $(\lambda-\theta,\nu")$ satisfies again the conditions (4.18). Hence a finite number of steps lead to the case $\lambda-\theta \leq \mu+\mu'$. Slights modifications of the three steps sketched above give then $(R_{\mu+\mu',-1})$.

4.4 Principal symbol of the solution

As already pointed out in the introduction, Theorem 4.3.3 improves the regularity results of J.M. Bony and Wang Weike. Now we shall go further: near each point of $\Sigma_j \setminus \Gamma$ $(1 \leq j \leq N)$ u is a single conormal wave belonging to $I^{\mu}(\Sigma_1)$ if $j=1$, to $I^{\mu'}(\Sigma_2)$ if $j=2$, to $I^{\mu+\mu'}(\Sigma_j)$ if $j \geq 3$, and we want to study the corresponding principal symbol of u (see 1.3 of chap.1). As usually, to get invariant formulation of the results, we consider u, Lu and $f(x,u)$ as half densities sections on Ω. We begin with de description of the principal symbols of the singularities on $\Sigma_1 \setminus \Gamma, \Sigma_2 \setminus \Gamma$. The results is quite similar to the corresponding one in Theorem 3.3.7. Recall that the δ-principal symbol and the notation $\delta(\mu)$ were defined in (3.3). We set $\delta = \delta(\mu)$.
We shall sometimes use the following properties, which follow easily from $(P_{4.16})$ in the case $\nu = -\infty$, that is from the more elementary case of a single hypersurface (see Piriou A. [P]).

$(\mathbf{P_{4.19}})$*If σ denotes the δ-principal symbol, then*

$$\sigma_{(uv)} = u_{|\Sigma_j}\sigma_v + v_{|\Sigma_j}\sigma_u$$

$$\sigma_{(F(u))} = F'(u)_{|\Sigma_j}\sigma_u$$

when $u,\ v \in I^{\mu}(\Sigma_j),\ \mu < -1,\ F \in C^{\infty}$.

Theorem 4.4.1 *In the conditions of Theorem 4.3.3, the* δ*-principal symbol of* u *near the points of* $C_j \backslash \{x_0\},\ j = 1, 2$, *is represented by* $\sigma_j e_j$, *where the (scalar) half- density* σ_j *satisfies across* Γ *the transport equation*

$$(i^{-1}\mathcal{L}_{H_j} + h_j)\sigma_j = f_j'\sigma_j$$

$\mathcal{L}_{H_j}$ *is the lie derivative with respect to the hamiltonian field* H_j *of* $L_{j,j}$, h_j *is the subprincipal symbol of* $L_{j,j}$, *and* $f_j' =< e_j', f_u'(.,u)e_j >_{|\Sigma_j}$ *(see the notations used for* $(4.4)_j$ *).*

More concretely consider for instance the case $j = 1$. Choose local coordinates such that $\Sigma_1 = \{x_1 = 0\}$, $\Gamma = \{x_1 = x_2 = 0\}$, $L_{1,1} = \partial_{x_2} + b$ with $b \in C^{\infty}$, $C_1^- \subset \{x_2 < 0\}$. Suppose that, near a point $x_0^- = (0, \alpha, x_0'')$ such that $\alpha < 0$ sufficiently small, u_1 satisfies

$$u_1(x) \equiv \int e^{ix_1\xi_1}c^-(x', \xi_1)d\xi_1 \ \ mod.I^{\mu-\delta}(\Sigma_1) \tag{4.20}$$

where $x' = (x_2, x'')$, $x'' = (x_3, .., x_n)$, $c^- \in S^{\mu}$.
Let $c(x', \xi_1)$ be the solution of the transport equation

$$(\partial_{x_2} + b(0, x'))c(x', \xi_1) = \frac{\partial f_1}{\partial u_1}(x, u)_{|x_1=0}c(x', \xi_1) \tag{4.21}$$

such that $c = c^-$ when $x_2 = \alpha$.

Remark that $\frac{\partial f_1}{\partial u_1}(x, u) \in I^{\mu,\mu'}(\Sigma_1, \Gamma) + I^{\mu',\mu}(\Sigma_2, \Gamma) + I^{\mu+\mu',-1}(\Sigma_{1,2}')$ thanks to Theorem 4.3.3 and results of section 4.2. Thus $(P_{4.3})$ shows that $\frac{\partial f}{\partial u_1}(x, u)_{|\Sigma_1} \in I^{\mu'}(\Gamma)$ and c is of the form

$$c(x', \xi_1) = K(x')c^-(\alpha, x'', \xi_1), \ \ K \in I^{\mu'-1}(\{x_2 = 0\}) \tag{4.22}$$

Therefore c is an ordinary symbol in S^{μ} when $x_2 \neq 0$; what means Theorem 4.4.1 is that $u_1(x) \equiv \int e^{ix\xi_1}c(x', \xi_1)d\xi_1$ *mod.* $I^{\mu-\delta}(\Sigma_1)$ if $x_2 \neq 0$ (in fact this is already knew from Theorem 3.2.1 when $x_2 < 0$, and the important point is now the case $x_2 > 0$).

For proving this result, we shall establish a more precise statement involving the behaviour of u_1, u_2 at the wedge Γ.

Proposition 4.4.2 *If $j=1$ or 2, u_j admits a decomposition of the form*

$$u_j = a_j U_j + r_j \quad with \tag{4.23}$$

$$a_j \in I^{\mu-1,\mu'-1}(\Sigma_1,\Gamma) + I^{\mu'-1,\mu-1}(\Sigma_2,\Gamma) + I^{\mu+\mu'-1,-1}(\Sigma'_{1,2})$$
$$U_1 \in I^{\mu}(\Sigma_1),\ U_2 \in I^{\mu'}(\Sigma_2),\ U_j = 0 \ on\ \Sigma_j$$
$$r_1 \in I^{\mu-\delta,\mu'-1+\delta}(\Sigma_1,\Gamma) + I^{\mu'-1,\mu}(\Sigma_2,\Gamma) + I^{\mu+\mu'-1,-1}(\Sigma'_{1,2})$$
$$r_2 \in I^{\mu-1,\mu'}(\Sigma_1,\Gamma) + I^{\mu'-\delta,\mu-1+\delta}(\Sigma_2,\Gamma) + I^{\mu+\mu'-1,-1}(\Sigma'_{1,2})$$

Proof of Proposition 4.4.2. and Theorem 4.4.1.
We have $L_{1,1}u_1 = f_1(x,u) - \sum_{j\geq 2} L_{1,j}u_j$. Like in the first step of the proof of Proposition 4.3.2, one can write $u_1 = \psi_1 + v_1$ with $\psi_1 \in I^{\mu'-1,\mu}(\Sigma_1,\Gamma) + I^{\mu+\mu'-1,-1}(\Sigma'_{1,2})$; $v_1 \in I^{\mu,\mu'-1}(\Sigma_1,\Gamma)$, $v_1 = 0$ on Σ_1.
A similar linearization gives now

$$L_{1,1}u_1 = \frac{\partial f_1}{\partial u_1}(x,\psi_1,u_2,..,u_N)u_1 + F_1, \tag{4.24}$$

$$F_1 \in I^{\mu-\delta,\mu'}(\Sigma_1,\Gamma) + I^{\mu',\mu}(\Sigma_2,\Gamma) + I^{\mu+\mu',-1}(\Sigma'_{1,2}).$$

Consider $U_1 \in I^{\mu}(\Sigma_1)$ defined by $U_1(x) = u_1(x_1,\alpha,x'')$ and set $\varpi_1 = \frac{\partial f_1}{\partial u_1}(x,\psi_1,u_2,..,u_N) - b$, the local coordinates beeing choosen as for (4.20).
Then the solution u_1 of (4.20) is given by the formula (4.14) corresponding to $j=1$. If we set $a_1 = e^{G_1}$ and

$$r_1(x) = e^{G_1(x)} \int (e^{-G_1}F_1)(x_1,s,x'')ds,$$

we get the desired decomposition (4.23) for $j=1$, except the condition $U_1 = 0$ on Σ_1. But it is easy to insure this condition since $U_{1|\Sigma_1} \in C^\infty$. Moreover, if u_1 is given by (4.20) in a neighbourhood of x_0^-, then $U_1(x) \equiv \int e^{ix_1\xi_1}c^-(\alpha,x'',\xi_1)d\xi \ mod I^{\mu-\delta}(\Sigma_1)$.
Since $a_1 \in I^{\mu-1}(\Sigma_1)$ and $r_1 \in I^{\mu-\delta}(\Sigma_1)$ close to the points of $\Sigma_1 \setminus \Gamma$, (4.23) and ($P_{4.19}$) show that $u_1(x) \equiv \int e^{ix_1\xi_1}\widetilde{c}(x',\xi_1)d\xi_1 \ mod\ ; I^{\mu-\delta}(\Sigma_1)$ close to the points of $\Sigma_1 \setminus \Gamma$, where $\widetilde{c}(x',\xi_1) = a_1(0,x')c^-(\alpha,x'',\xi_1)$. Since $\widetilde{c}$ satisfies the transport equation (4.21) and $\widetilde{c} = c^-$ if $x_2 = \alpha$, we get $\widetilde{c} = c$. Thus the assertions of Proposition 4.4.2 and Theorem 4.4.1 are proven when $j=1$. The case $j=2$ is similar.

We now choose j such that $3 \leq j \leq N$ and we study the principal symbol of "nonlinear" singularity of u on Σ_j. An intrinsic result is given in Bougrini H., Piriou A., Varenne J.P. [BPV]. Here we restrict to the formulation of this result when local coordinates are choosen such that

$$\Sigma_j = \{x_1 = 0\} \ , \ \Gamma = \{x_1 = x_2 = 0\} \ , \ L_{j,j} = \partial_{x_2} + b \tag{4.25}$$

with $b \in C^\infty$, $C_j^- \subset \{x_2 < 0\}$, $\{x_2 = 0\}$ is non characteristic. If $k = 1$ or 2, we set $\Sigma_k = \{x_1 = \gamma_k(x')\}$.
Theorem 4.4.1 and an easy change of variable in (4.22) show that, close to the points of $\Sigma_k \setminus \Gamma$:

$$u_k(x) \equiv \int e^{i(x_1 - \gamma_k(x'))\xi_1} c_k(x', \xi_1) d\xi_1 \tag{4.26}$$

mod.$I^{\mu-\delta}(\Sigma_1)$ or $I^{\mu'-\delta}(\Sigma_2)$ if $k = 1$ or $k = 2$, where c_k is of the form $c_k(x', \xi_1) = K_k(x')\tilde{c}_k(x', \xi_1)$, $\tilde{c}_k \in S^\mu$ or $S^{\mu'}$, $K_k \in I^{\mu'-1}(\{x_2 = 0\})$ or $I^{\mu-1}(\{x_2 = 0\})$.
Thus the value $c_k(o, x'', \xi_1)$ of c_k for $x_2 = 0$ is an ordinary symbol in S^μ or $S^{\mu'}$, which is determined from the past c_k^- by the corresponding transport equation of Theorem 4.4.1. Remark also (see Proposition 4.4.2 and its proof) that (4.26) makes sense for $x_2 = 0$: $c_k(0, x'', \xi_1)$ is the δ-principal symbol of the trace $u_{k|x_2=0}$. Now we can state the main result of this section.

Theorem 4.4.3 *In the conditions of Theorem 4.3.3, let j be fixed such that $3 \leq j \leq N$, Consider local coordinates (4.25) and the above notations γ_k, c_k $(k = 1, 2)$. Then*

$$U_j(x) \equiv \int e^{ix_1\xi_1} c_j(x', \xi_1) d\xi_1 \ mod \ I^{\mu+\mu'-\delta}(\Sigma_j) \tag{4.27}$$

close to the points of C_j^+, where c_j is the symbol of degree $\mu + \mu'$ defined by the transport equation

$$(\partial_{x_2} + b)c_j = f_j'(0, x')c_j \quad (x_2 \geq 0) \tag{4.28}$$

and the initial condition

$$c_j(0, x'', \xi_1) = \frac{2\pi}{|\lambda_1 - \lambda_2|} f_j''(0, 0, x'') c_1(0, x'', \mu_1\xi_1) c_2(0, x'', \mu_2\xi_1) \tag{4.29}$$

We have set
$f_j'(x) = \dfrac{\partial f_j}{\partial u_j}(x, u_1, .., u_N)$

$$f_j^{"}(x) = \frac{\partial^2 f_j}{\partial u_1, \partial u_2}(x, u_1, .., u_N)$$
$$\lambda_k = (\partial_{x_2}\gamma_k)(0, x^{"}), \quad k = 1, 2 \ ; \mu_1 = \frac{\lambda_2}{\lambda_2 - \lambda_1}, \ \mu_2 = \frac{\lambda_1}{\lambda_1 - \lambda_2}$$

Proof. In order to simplify the presentation we shall detail only a particular case which contains the main difficulties of the question. We refer to Bougrini H., Piriou A., Varenne J.P. [BPV] for the general case. We assume the following additional hypothesis: $s > 3/2$ (thus $\mu < -2, \delta = 1$, and the δ-principal symbol is the ordinary principal symbol), the size of the system is $N = 3$ (thus we have only $\Sigma_1, \Sigma_2, \Sigma_3$) and we study the principal symbol of u_3), the number of variables is $n = 2$ (thus the reduction (2.3.2) is now a trivial diagonalisation, and $L_{j,k} = 0$ when $j \neq k$ in (4.4)j).
We begin with a linearisation of $(4.4)_3$ similar to the one which was used in the first step of the proof of Proposition 4.3.2. The local coordinates are choosen such that: $\Sigma_3 = \{x_1 = 0\}$, $\Gamma = \{x_1 = x_2 = 0\}$, $C_3^- \subset \{x_2 < 0\}$, $L_{3,3} = \partial_{x_2} + b \ \ (b \in C^\infty)$.
Changing $f_3(x, u)$ into $f_3(x, u) - bu$ and u_3 into $u_3 - u_3(0)$ allow to assume that $L_{3,3} = \partial_{x_2}$ and $u_3(0) = 0$. We can write $u_3 = \psi_3 + v_3$ with $\psi_3 \in I^{\mu-1,\mu'}(\Sigma_1, \Gamma) + I^{\mu'-1,\mu}(\Sigma_2, \Gamma)$, $v_3 \in I^{\mu+\mu',-1}(\Sigma_3, \Gamma)$, $v_3 = 0$ on Σ_3.
We set $\gamma = (u_1, u_2, \psi_3)$. Taylor's formula gives

$$\partial_{x_2} u_3 - \varphi_3 u_3 = Y_1 + Y_2 \ , \tag{4.30}$$

$\varphi = \frac{\partial f_3}{\partial u_3}(x, \gamma)$, $Y_1 = f_3(x, \gamma) - \frac{\partial f_3}{\partial u_3}(x, \gamma)\psi_3$, $Y_2 = v_3^2 F(x, u, \psi_3)$,
$F \in C^\infty$, and we deduce that

$$u_3 = \Phi + Z_1 + Z_2 \ , \tag{4.31}$$

where $\Phi(x) = u_3(x_1, \alpha)e^{G(x)}$, $G(x) = \int_\alpha^{x_2} \varphi_3(x_1, s)ds$,
and $Z_l = e^G \int_\alpha^{x_2}(e^{-G}Y_l)(x_1, s)ds$, $l = 1, 2$. Here $\alpha < 0$ is choosen sufficiently close to 0.

Our goal is to study the principal symbol of u_3 on $\Sigma_3^+ = \Sigma_3 \cap \{x_2 > 0\}$. We shall denote by $\sigma(v)$ the principal symbol of an element v in $I^{\mu+\mu'}(\Sigma_3^+)$, and the notation $v \sim w$ will stand for $\sigma(v) = \sigma(w)$.
We define the following algebras (which are also stable under the action of smooth functions)

$$\left\{\begin{array}{lcl} \mathcal{A} & = & I^{\mu,\mu'}(\Sigma_1,\Gamma) + I^{\mu',\mu}(\Sigma_2,\Gamma) + I^{\mu+\mu'-1,-1}(\Sigma_3,\Gamma) \\ \mathcal{B} & = & I^{\mu,\mu'}(\Sigma_1,\Gamma) + I^{\mu',\mu}(\Sigma_2,\Gamma) + I^{\mu+\mu',-1}(\Sigma_3,\Gamma) \\ \mathcal{C} & = & I^{\mu-1,\mu'}(\Sigma_1,\Gamma) + I^{\mu'-1,\mu}(\Sigma_2,\Gamma) + I^{\mu+\mu',-1}(\Sigma_3,\Gamma) \\ \mathcal{D} & = & I^{\mu-1,\mu'}(\Sigma_1,\Gamma) + I^{\mu'-1,\mu}(\Sigma_2,\Gamma) + I^{\mu+\mu'-1,-1}(\Sigma_3,\Gamma) \end{array}\right. \tag{4.32}$$

We consider also the space

$$E = I^{\mu,\mu'-1}(\Sigma_1,\Gamma) + I^{\mu',\mu-1}(\Sigma_2,\Gamma) + I^{\mu+\mu'-1,-1}(\Sigma_3,\Gamma) \tag{4.33}$$

Note that $(P_{4.11})$ shows that

$$g \in E \Rightarrow \int_\alpha^{x_2} g(x_1,s)ds \sim 0 \tag{4.34}$$

We now return to (4.30), (4.31)

First step: $Z_2 \sim 0$.
Since $u, \psi_3 \in \mathcal{B}$, we get $F(x,u,\psi_3) \in \mathcal{B}$.
Thanks to $(P_{4.8})$ we can write $v_3 = x_1 w_3$ where $w_3 \in I^{\mu+\mu'+1,-1}(\Sigma_3,\Gamma)$, Thus $Y_2 = x_1^2 Y_2'$ when $Y_2' = w_3^2 F(x,u,\psi_3)$ belongs to the algebra $I^{\mu,\mu'}(\Sigma_1,\Gamma)+ I^{\mu',\mu}(\Sigma_2,\Gamma) + I^{\mu+\mu'+1,-1}(\Sigma_3,\Gamma)$. Using $(P_{4.10})$ we deduce that $Y_2 \in E$, and (4.34) gives $Z_2 \sim 0$.

Second step: Expression of $\sigma(u_3)$.
The first step and (4.31) show that

$$u_3(x) \sim e^{G(x)}(u_3(x,\alpha) + \psi(x)) \tag{4.35}$$

where

$$\psi(x) = \int_\alpha^{x_2} (e^{-G}Y_1)(x_1,s)ds. \tag{4.36}$$

Since $\gamma, \psi_3 \in \mathcal{A}$, we get $\varphi_3, Y_1 \in \mathcal{A}$, $G, e^{\pm G} \in \mathcal{C}$, $e^{-G}Y_1 \in \mathcal{B}$, $\psi \in \mathcal{C}$.
We can write

$$G = P + R, \tag{4.37}$$

$R \in I^{\mu-1,\mu'}(\Sigma_1,\gamma) + I^{\mu'-1,\mu}(\Sigma_2,\Gamma)$, $P \in I^{\mu+\mu',-1}(\Sigma_3,\Gamma)$, $P = 0$ on Σ_3, $P(x) = R(x) = 0$ if $x_2 = \alpha$ (the last condition follows from the nullity of

$G(x)$ if $x_2 = \alpha$).
By Taylor's formula we get

$$e^{-G} = e^{-R}e^{-P} = e^{-R}(1 - P + P^2 F_1(P)), \tag{4.38}$$

with $F_1 \in C^\infty$.
Since $P^2 \in I^{\mu+\mu'-1,-1+\varepsilon}(\Sigma_3, \Gamma)$ thanks to $(P_{4.16})$ and vanishes on Σ_3, $(P_{4.8})$ shows that $P^2 = x_1 Q$ with $Q \in I^{\mu+\mu',-1+\varepsilon}(\Sigma_3, \Gamma)$. Thus $e^{-R}P^2F_1(P)Y_1 = x_1 S$, where $S = e^{-R}QF_1(P)Y_1$ is in the algebra $I^{\mu,\mu'}(\Sigma_1, \Gamma)$-$I^{\mu',\mu}(\Sigma_2, \Gamma)$+$I^{\mu+\mu',-1+\varepsilon}(\Sigma_3, \Gamma)$, and $(P_{4.10})$, $(P_{4.11})$ show that $\int_\alpha^{x^2} x_1 S(x_1, s)ds$ 0.
It follows (see (4.36), (4.38)) that

$$\psi \sim A - B,\ A(x) = \int_\alpha^{x_2} e^{-R} Y_1 ds,\ B(x) = \int_\alpha^{x_2} e^{-R} P Y_1 ds \tag{4.39}$$

An integration by parts gives

$$B(x) = (AP)(x) - \int_\alpha^{x_2} (A\frac{\partial P}{\partial x_2})(x_1, s))ds \tag{4.40}$$

Using $(P_{4.19})$ and remarking that $u_3(., \alpha)$ is smooth in a neighbourhood of 0, we deduce from (4.35), (4.39) that

$$\sigma(u_3) = e^{G(0,x_2)}\{(u_3(0, \alpha) + A(0, x_2))\sigma(G) + \sigma(A) - \sigma(B)\} \tag{4.41}$$

(remark that $\psi = A$ on Σ_1 since $G = R$ on Σ_1)

Third step: study of $\sigma(B)$

We shall prove that A can be replaced by $A(0)$ in the integrand of (4.40). Indeed, since $A \in C$, we can write

$$A = A' + A'', \tag{4.42}$$

with $A' \in I^{\mu-1,\mu'}(\Sigma_1, \Gamma) + I^{\mu'-1,\mu}(\Sigma_2, \Gamma)$, $A'' \in I^{\mu+\mu',-1}(\Sigma_3, \Gamma)$, $A'' = 0$ on Σ_3.

Writing $A'' = x_1 A''_1$ with $A''_1 \in I^{\mu+\mu'+1,-1}(\Sigma_3, \Gamma)$ and noticing that $\dfrac{\partial P}{\partial x_2} \in I^{\mu+\mu',0}(\Sigma_3, \Gamma)$ vanishes on Σ_3 (see (4.37)), we get $\dfrac{\partial P}{\partial x_2} = x_1 Q$ with

$Q \in I^{\mu+\mu'+1,0}(\Sigma_3,\Gamma)$. Thus $A"\dfrac{\partial P}{\partial x_2} = x_1^2 Q_1$, $Q_1 \in I^{\mu+\mu'+1,0}(\Sigma_3,\Gamma)$ and $\int_\alpha^{x_2} A"\dfrac{\partial P}{\partial x_2} ds \sim 0$ since it belongs to $I^{\mu+\mu'-1,-1}(\Sigma_3,\Gamma)$.

We consider now the contribution of A'. We can write $A' = A'(0) + x_1A_1 + x_2A_2$, with $A_1, A_2 \in I^{\mu,\mu'}(\Sigma_1,\Gamma) + I^{\mu',\mu}(\Sigma_2,\Gamma)$. Since $A_1\dfrac{\partial P}{\partial x_2}$ belong to the algebra $I^{\mu,\mu'}(\Sigma_1,\Gamma)I^{\mu',\mu}(\Sigma_2,\Gamma) + I^{\mu+\mu',0}(\Sigma_3,\Gamma)$, we get $\int_\alpha^{x_2} x_1\, A_1\dfrac{\partial P}{\partial x_2} ds \sim 0$.
For A_2, we remark that $\dfrac{\partial P}{\partial x_2} = \varphi_3 - \dfrac{\partial R}{\partial x_2}$ (see (4.37), (4.31)), thus $\dfrac{\partial P}{\partial x_2} \in \mathcal{A}$, $A_2\dfrac{\partial P}{\partial x_2} \in \mathcal{A}$, $x_2A_2\dfrac{\partial P}{\partial x_2} \in E$, thus (see (4.34)) we have proved that A can be replaced by $A'(0)$ in the integrand of (4.40). But $A'(0) = A(0)$ (see (4.37)) and $P = 0$ on σ_3 (see (4.37)), thus it follows from $(P_{4.19})$ that $B \sim AP - A(0)P$, $\sigma(B) = (A(0,x_2) - A(0,0))\sigma(G)$.
From (4.41) we thus obtain $\sigma(u_3) = e^{G(0,x_2)}\sigma(A) + \varepsilon(\alpha)\sigma(G)$ where $\varepsilon(\alpha) = u_3(0,\alpha) + A(0,0)$
$= u_3(0,\alpha) + \int_\alpha^0 (e^{-G}Y_1)(0,s)ds = 0$,
(see (4.31), and remark that $u_3(0,0) = 0$, $Z_2 = 0$ on Σ_1).
We have proved that

$$\sigma(u_3) = e^{G(0,x_2)}\sigma(A) \tag{4.43}$$

Forth step: Study of $\upsilon(A)$.

Lemma 4.4.4 *Let u_1, u_2 be decomposed as in Proposition 4.4.2. Let r be vector valued with components in $I^{\mu-1,\mu'}(\Sigma_1,\Gamma) + I^{\mu'-1,\mu}(\Sigma_2,\Gamma)$.*
If F is smooth function we set

$$F(u_1,u_2,r) = g\ ,\ c = \frac{\partial^2 F}{\partial x_1, \partial x_2}(x_1,x_2,r)a_1a_2$$

Then:

i) $g \equiv c(0,0)u_1u_2$ *mod. E (see (4.33))*

ii) $\int_\alpha^{x_2} g(x_1,s)ds \sim c(0,0)\int_\alpha^{x_2}(U_1U_2)(x_1,s)ds$

Proof. We set $U = (u_1,u_2,r)$, $\rho = (r_1,r_2,r)$, $V = (a_1U_1, a_2U_2)$.
Taylor's formula yields

$$F(U) = F(\rho) + \sum_{j=1,2}\frac{\partial F}{\partial u_j}(\rho)a_jU_j +$$

$$\sum_{|\alpha|=2} \frac{1}{\alpha!}(\partial^\alpha F)(\rho)V^\alpha + \sum_{|\alpha|=3} V^\alpha F_\alpha(U,\rho), \quad \text{with } F_\alpha \in C^\infty.$$

Since $R,\ a_1,\ a_2 \in \mathcal{D}$ (see (4.32), we get $\frac{\partial F}{\partial u_j}(\rho)a_j \in \mathcal{D}$. Thus $\frac{\partial F}{\partial u_1}(\rho)a_1U_1$ belongs to the algebra $I^{\mu,\mu'-1}(\Sigma_1,\Gamma)+I^{\mu'-1,\mu}(\Sigma_2,\Gamma)+I^{\mu+\mu'-1,-1}(\Sigma_3,\Gamma)$ since $U_1 \in I^\mu(\Sigma_1) = I^{\mu,-\infty}(\Sigma_1,\Gamma)$, and this algebra is included in E.
In the same way we obtain $\frac{\partial F}{\partial u_2}(\rho)a_2U_2 \in E$, $\partial^\alpha F(\rho)V^\alpha \in E$ if $\alpha = (2,0)$ or $\alpha = (0,2)$.
Consider now the termes with $|\alpha| = 3$. We denote $(a_1, a_2) = a$ and we use local coordinates such that $\Sigma_j = \{x_j = 0\}$, $j = 1,2$
Since $U,\ \rho,\ a \in \mathcal{A}$, we have $F_\alpha(U,\rho)a^\alpha \in \mathcal{A}$.
If $\alpha = (\alpha_1,\alpha_2)$ is sucht that $\alpha_1 \geq 2$, then $U_1^{\alpha_1} \in I^{\mu-1}(\Sigma_1)$ and vanishes ou Σ_1. It follows that $U_1^{\alpha_1} = x_1W_1$ with $W_1 \in I^\mu(\Sigma_1)$, and $F_\alpha(U,\rho)V^\alpha \in x_1\mathcal{A} \subset E$. The case $\alpha_2 \geq 2$ is similar.
Therefore $F(u_1,u_2,\rho) \equiv cU_1U_2 \ mod\ E$, with $c = \frac{\partial^2 F}{\partial u_1 \partial u_2}(\rho)a_1a_2 \in \mathcal{D}$.
That c can be replaced by $c(0)$ is proven like in the third step. Thus the part i) is established, and the part ii) follows from (4.34).

Returning to the definition of A (see (4.39), Y_1 (see (4.30)) and R (see (4.37), setting $\rho = (R,\psi_3)$ and recalling that $\psi_3(0) = U_3(0) = 0$, Lemma 4.4.4 gives $e^{-R}Y_1 \equiv e^{-G(0)}f_3''(0)a_1(0)a_2(0)U_1U_2 \ mod\ E$ and

$$A \equiv e^{-G(0)}f_3''(0)a_1(0)a_2(0)\int_\alpha^{x_2}(U_1U_2)(x_1,s)ds\ , \tag{4.44}$$

where we have used the notations of Theorem 4.4.3

Fifth step: study of $W = \int_\alpha^{x_2} U_1U_2 ds$.

Using again the notations of Theorem 4.4.3, U_1 and U_2 are of the form $U_k(x) = \int e^{i(x_1-\gamma_k(x_2))\eta}b_k(x,\eta)d\eta,\ \ k = 1,2$
with $b_1 \in S^\mu,\ b_2 \in S^{\mu'}$.
Therefore

$$U_1U_2(x) = \int e^{i(x_1-\gamma_1(x_2))\eta_1+i(x_1-\gamma_2(x_2)))\eta_2}b_1(x,\eta_1)b_2(x,\eta_2)d\eta_1d\eta_2.$$

Since $\gamma_k(0) = 0$, we have $\gamma_k(x_2) = x_2\delta_k(x_2)$, with $\delta_k(0) = \lambda_k \neq 0 \quad (k = 1,2)$.

The change of variables $\xi_1 = \eta_1 + \eta_2$, $\xi_2 = -\delta_1(x_2)\eta_1 - \delta_2(x_2)\eta_2$ gives

$$U_1U_2(x) = \int e^{i(x_1\xi_1 + x_2\xi_2)} a(x, \xi_1, \xi_2) d\xi_1 d\xi_2$$

where

$$a(x, \xi_1, \xi_2) = \frac{1}{|\delta_1 - \delta_2|} b_1(x, \frac{\xi_2 + \delta_2\xi_1}{\delta_2 - \delta_1}) b_2(x, \frac{\xi_2 + \delta_1\xi_1}{\delta_1 - \delta_2})$$

Taylor's formula proves easily that
$W \sim V$ if $V(x) = \int_\alpha^{x_2} v(x_1, s)ds$, $v(x) = \int e^{i(x_1\xi_1 + x_2\xi_2)} a(0, \xi_1, \xi_2) d\xi_1 d\xi_2$.
Then a simple adaptation of the proof of $(P_{4.11})$ gives

$$V(x) \sim 2\pi \int e^{ix_1\xi_1} a(0, \xi_1, 0) d\xi_1$$

Finally (4.43), (4.44) show that

$$\sigma(u_3) = e^{S(x_2)} f_3''(0) a_1(0) a_2(0) \frac{1}{|\delta_1 - \delta_2|} b_1(0, \mu_1\xi_1) b_2(0, \mu_2\xi_1)$$

where $S(x_2) = G(0, x_2) - G(0, 0)$.
The conclusion of Theorem 4.4.3 follows since $a_k b_k = c_k$ $(k = 1, 2)$, and since the function $e^{S(x_2)}$ satisfies the transport equation (4.28) and the initial condition $e^{S(0)} = 1$.

4.5 Examples

At the δ-principal symbol level, Theorem 4.3.3 allows a precise description of the singularities after interaction in term of the two incident singularities. In order to shorten the formulation, we shall apply Theorem 4.3.3 when $\Sigma_1 = \{x_1 = x_2\}$, $\Sigma_2 = \{x_1 = -x_2\}$, the general case being similar. We study the singularities on $\Sigma_j = \{x_1 = 0\}$ close to the origin, assuming that $\frac{\partial^2 f_j}{\partial u_1 \partial u_2}(0, u(0)) \neq 0$. We consider the case of two incident singularities which are of classical conormal type (without logarithms, see Examples 3.2.5) at the δ-principal symbol level. Suppose for instance (the generalization is straightforward) that u_1, u_2 have respectively incident δ-principal parts of the form $K(x)\mathcal{H}(x_2 - x_1)(x_2 - x_1)^{\alpha - 1}$, $K(x)\mathcal{H}(x_2 + x_1)(x_2 + x_1)^{\beta - 1}$, with $\alpha, \beta \in \mathbb{C}$, $\Re\alpha$ and $\Re\beta > 1$; here and after the notation K stands for diverse non vanishing smooth functions. Formula (4.29) gives

$$c_j(0,x^{"},\xi_1)=\pi f_j^{"}(0,0,x^{"})c_1(0,x^{"},\frac{\xi_1}{2})c_2(0,x^{"},\frac{\xi_1}{2}).$$

Since the Fourier transform of the distribution $\mathcal{H}(\pm x_1)(\pm x_1)^{\lambda-1}/\Gamma(\lambda)$ is $e^{\mp i\lambda\frac{\pi}{2}}(\xi_1\mp i0)^{-\lambda}$ when $\Re\lambda>0$, we obtain ; for $|\xi_1|$ large :
$c_1((0,x^{"},\frac{\xi_1}{2})\sim K(x^{"})(\xi_1+i0)^{-\alpha}$, $c_2((0,x^{"},\frac{\xi_1}{2})\sim K(x^{"})(\xi_1-i0)^{-\beta}$, $c_j((0,x^{"},\xi_1$
$K(x^{"})(\mathcal{H}(\xi_1)\xi_1^{-\alpha-\beta}+e^{i\pi(\beta-\alpha)}\mathcal{H}(-\xi_1)(-\xi_1)^{-\alpha-\beta})$.

Case $\alpha,\beta\in I\!N$. (This case correponds to piecewise smooth principal parts in the two incoming waves). Since $e^{i\pi(\beta-\alpha)}=(-1)^{\beta+\alpha}$, we have $c_j((0,x^{"},\xi_1)=c(x^{"})\xi_1^{-\alpha-\beta}$ for large $|\xi_1|$, and integration of the transport equation (4.28) gives $u_j\sim K(x^{'})\mathcal{H}(x_1)x_1^{\alpha+\beta-1}$. Therefore u_j admits also a principal part of piecewise smooth type.

Case $\alpha,\beta\notin I\!N$, $\alpha+\beta\notin I\!N$. Then (1.35) shows that

$$u_j\sim K(x^{'})(A_+\mathcal{H}(x_1)x_1^{\alpha+\beta-1}+A_-\mathcal{H}(-x_1)(-x_1)^{\alpha+\beta-1})$$

with $A_+=sin(\pi\beta)$ and $A_-=sin(\pi\alpha)$. Therefore u_j admits also a principal part of classical (without logarithms) type.

Case $\alpha,\beta\notin I\!N$, $\alpha+\beta\in I\!N$. We set $\alpha+\beta=m$. From (1.36) we get

$$u_j\sim K(x^{'})(A\mathcal{H}(x_1)x_1^{m-1}/(m-1)!\ -B(ix_1)^{m-1}ln|x_1|)$$

with $A=\pi i^m(1+(-1)^m e^{i\pi(\beta-\alpha)})$, $B=1-(-1)^m e^{i\pi(\beta-\alpha)}$.
Remark that $B\neq 0$ (indeed $B=0$ means $e^{i\pi(\beta-\alpha)}=(-1)^m$, that is $\beta-\alpha\equiv\beta+\alpha$ (2), which is impossible since $\alpha\notin Z\!\!Z$), thus a logarithmic singularity appears in the principal part of u_j. For instance if the incident waves u_1, u_2 are respectively of the form $K(x)\mathcal{H}(x_2-x_1)(x_2-x_1)^{\frac{1}{2}}$, $K(x)\mathcal{H}(x_2+x_1)(x_2+x_1)^{\frac{1}{2}}$, we have $\alpha=\beta=3/2$, $\alpha+\beta=m=3$, $A=0$, $B=2$, $u_j\sim K(x^{'})x_1^2ln|x_1|$.

Chapter 5

Symbolic calculus: interaction of piecewise smooth waves.

G. LASCHON

5.1 Introduction

The chapters 3 and 4 were devoted to the symbolic study of the principal conormal singularities which appear by the interaction of two transversal conormal waves. In this chapter we are interested in the particular case where the two interacting waves are piecewise smooth (see chapter 1 section 3).

In this way let's consider a solution u of the system (2.1) under the assumptions of Theorem 2.1.2, and suppose

(H) $u_{|\Omega^-} \in pC^\infty_{\Sigma_1} + pC^\infty_{\Sigma_2}$, where $pC^\infty_{\Sigma_i}$ is the space of the piecewise smooth functions with respect to Σ_i.

Our aim is to state precisely the kind of the singularities after interaction near the points of Σ_i^+ for $1 \leq i \leq r$.

An element of the answer is given by G. Métivier and J. Rauch in $[MR2]$ which worked under the assumption of strict hyperbolicity for the system with respect to a timelike function. They established that if the edge $\Gamma = \Sigma_1 \cap \Sigma_2$ is included in a spacelike hypersurface then a solution u verifying (H) in the past is still piecewise smooth with respect to Σ

after interaction. Let's remark that under such geometrical hypothesis the number r of the characteristic hypersurfaces passing through Γ is equal to the size N of the system. In fact this equality is essential as G. Métivier and J. Rauch showed with a counter-example in $[MR1]$ where a non piecewise smooth singularity appears after the interaction.

In this chapter we propose a symbolic study of the singularities showing the microlocal mechanism of the creation and the propagation of non piecewise smooth type singularities. We assume that $r < N$. We will see, by a significant example, that the microlocally elliptic part of the system originates the apparition of a non piecewise smooth singularity on the edge Γ which can propagate along the charateristic hypersurfaces $\Sigma_1, \ldots, \Sigma_r$ if the nonlinearity of the system is enough strong.

First, one note that the study of the principal symbols restricted to the case $r = N$ in the chapters 3 and 4 is still available if $2 \leq r < N$. Indeed the construction in the chapter 4 section 3 yields two families of independant vectors in $\mathbb{C}^N$ $(e_1(x), \ldots, e_r(x))$ and $(e'_1(x), \ldots, e'_r(x))$, regular with respect to x, where each vector $e_i(x)$ or $e'_i(x)$ for $1 \leq i \leq r$ generates the kernel of $l(x, \xi)$ or ${}^t l(x, \xi)$ when $(x, \xi) \in N(\Sigma_i)$. Considering the complex values for ξ which annuls *det* $l(x, \xi)$, one completes each family in order to obtain two basis of $\mathbb{C}^N$ such that the system (2.1) is equivalent to

$$\sum_{k=1}^{N} L_{j,k}(u_k) = f_j(x, u) \text{ for } 1 \leq j \leq N$$

where the principal part of the first order differential operator $L_{j,k}$ is tangent to Σ_j and Σ_k for $j \neq k$ and $1 \leq j, k \leq r$, $(L_{j,k})_{r+1 \leq j,k \leq N}$ is elliptic on a neighborhood of $N(\Gamma)$, the conormal space with respect to the edge Γ, and $u = \sum_{k=1}^{N} u_k e_k$ such that $u_1, \ldots, u_r$ are real scalar functions. The principal part of the $L_{j,j}$ for $1 \leq j \leq r$ is proportionnal to the projection of the corresponding hamiltonian field of L on $N(\Sigma_j)$.

Each principal singularity appears on the corresponding composant u_j of the solution. More precisely, adapting the demonstrations of the chapter 4, we obtain the same estimations as in Theorem 4.3.3 for $j = 1, \ldots, r$ (replacing N by r), and for $r + 1 \leq j \leq N$ we have

$$u_j \in I^{\mu-1,\mu'}(\Sigma_1, \Gamma) + I^{\mu'-1,\mu}(\Sigma_2, \Gamma) + I^{\mu+\mu'-1,-1}(\Sigma'_{1,2,j}).$$

The Proposition 4.4.2 is still available and the symbols of the principal conormal singularities with respect to the $\Sigma_3^+, \ldots, \Sigma_r^+$ satisfy the transport equations and the initial conditions of the Theorem 4.4.3 (see [La] for details).

We saw in the chapter 1 the characterization of the symbol of a function piecewise smooth with respect to one hypersurface. In this way we can note that the principal singularities of the solution u verifying (H) is still piecewise smooth after the interaction as the transport equations satisfied by their symbols shows. So our aim is to go deeper into the aymptotic developments of the complete symbols to find the obstruction to the piecewise smooth propagation. We are going to treat just a significant example where $r = 2$ allowing us a more systematic calculus of the complete symbol of the solution.

The studied differential system is inspired by an example proposed by G. Métivier.

Let (u, v, w_1, w_2) the solution of the system on $\mathbb{R}^3$

$$\left.\begin{aligned} (\partial_t + \partial_x)u &= 0 \\ (\partial_t + \partial_y)v &= uw_1 \end{aligned}\right\} \tag{5.1}$$

$$\left.\begin{aligned} (\partial_t - \partial_x)w_1 + \partial_y w_2 &= \psi(t)uv \\ \partial_y w_1 + (\partial_t + \partial_x)w_2 &= 0 \end{aligned}\right\} \tag{5.2}$$

with $\psi(t) = \left| \begin{array}{l} 0 \text{ if } t < -\varepsilon \\ 1 \text{ if } t > -\frac{\varepsilon}{2} \end{array}\right.$ for a $\varepsilon > 0$, and verifying the following conditions in the past $\{t < -\varepsilon\}$:

$$\begin{aligned} &u(t, x, y) = y^{k_1 - 1}\mathcal{H}(y) \text{ with } k_1 \geq 2 \\ &v(t, x, y) = x^{k_2 - 1}\mathcal{H}(x) \text{ with } k_2 \geq 2 \\ &w_1 = w_2 = 0. \end{aligned} \tag{5.3}$$

Let's remind that $\mathcal{H}$ designates the Heaviside function and remark that the system is symmetric hyperbolic with respect to t.

In the past this solution has conormal singularities of piecewise smooth type with respect to the charateristic hypersurfaces $\Sigma_1 = \{y = 0\}$ and $\Sigma_2 = \{x = 0\}$ which are transverse along the edge $\Gamma = \{x = y = 0\}$. There is not any other characteristic hypersuface containing Γ. It expresses the fact that the part (5.2) of the system is microlocally elliptic on a neighborhood of $N(\Gamma)$.

From the conditions in the past, the general study described in the chapter 3 and previously extended to the case $r < N$ gives

$$\begin{array}{l} u \in I^{-k_1}(\Sigma_1) \\ v \in I^{-k_1-1,-k_2}(\Sigma_1,\Gamma) + I^{-k_2,-k_1-1}(\Sigma_2,\Gamma) \\ w_1, w_2 \in I^{-k_1-1,-k_2}(\Sigma_1,\Gamma) + I^{-k_2-1,-k_1}(\Sigma_2,\Gamma). \end{array} \tag{5.4}$$

We are going to prove the

Theorem 5.1.1 *On a neighborhood of 0, near each point of $\Sigma_2^+ = \Sigma_2 \cap \{y > 0\}$, we have*

$$v(t,x,y) = cx^{2k_1+k_2-1} ln|x| + v_1(t,x,y) + v_2(t,x,y)$$

where v_1 is piecewise smooth, v_2 is in $I^{-2k_1-k_2-1}(\Sigma_2)$, and $c \neq 0$.

One can find in $[La]$ a similar detailed example using three characteristic hypersurfaces and a 5×5 strictly hyperbolic system, inpired by $[MR1]$

5.2 Symbols and piecewise smooth regularity

By the change of variables $x_1 = y$, $x_2 = x$, $x_3 = t - x - y$, we obtain from (5.1) and (5.2)

$$\left.\begin{array}{r} \partial_{x_2} u = 0 \\ \partial_{x_1} v = uw_1 \end{array}\right\} \tag{5.1}$$

$$\left.\begin{array}{r} (-\partial_{x_2} + 2\partial_{x_3})w_1 + (\partial_{x_1} - \partial_{x_3})w_2 = \Psi(x_1+x_2+x_3)uv \\ (\partial_{x_1} - \partial_{x_3})w_1 + \partial_{x_2} w_2 = 0 \end{array}\right\} \tag{5.2}$$

with the new conditions in the past :

for $x_1 + x_2 + x_3 < -\varepsilon$,

$$\begin{array}{l} u = x_1^{k_1-1}\mathcal{H}(x_1) \text{ with } k_1 \geq 2 \\ v = x_2^{k_2-1}\mathcal{H}(x_2) \text{ with } k_2 \geq 2 \\ w_1 = w_2 = 0. \end{array} \tag{5.3}$$

Now we have $\Gamma = \{x_1 = x_2 = 0\}$, $\Sigma_1 = \{x_1 = 0\}$, $\Sigma_2 = \{x_2 = 0\}$, and for $j = 1, 2$

$$u_j = x_j^{k_j-1}\mathcal{H}(x_j) = U_j + g_j \tag{5.4}$$

with $g_j \in C^\infty$, $U_j = \displaystyle\int e^{ix_j\xi_j} c_j \frac{\chi(\xi_j)}{\xi_j^{k_j}} d\xi_j$, and $c_j = \frac{1}{2\pi}(-i)^{k_j}(k_j-1)!$.

The microlocal study of the conormal singularities which appear after the interaction necessitates the description of diverse class of symbols.

Symbols with one variable of frequency. One recalls that $S^{\mu}(I\!R^n \times I\!R)$ is the set of the symbols of order μ that is the C^{∞} functions $a(x, \xi)$ with $x \in I\!R^n$ and $\xi \in I\!R$, such that for $\alpha \in I\!N^n$, $\beta \in I\!N$, and $\mathcal{K}$ a compact of $I\!R^n$,

$$|\partial_x^{\alpha} \partial_{\xi}^{\beta} a(x, \xi)| \leq C(1 + |\xi|)^{\mu - \beta} \text{ for all } (x, \xi) \in \mathcal{K} \times I\!R.$$

For an hypersurface Σ and $\mu \in I\!R$, as in the chapter 3 section 2 we write $I^{\mu}(\Sigma) = H^{-\frac{1}{2} - \mu}(\Sigma)$ the space of conormal distributions of degree μ with respect to Σ that is the space of the distributions f smooth out of Σ and which can be written

$$f(x) = \int e^{ix_1 \xi_1} a(x, \xi_1) d\xi_1, \quad a \in S^{\mu}(I\!R^n \times I\!R) \tag{5.5}$$

near each point of Σ in coordinates such that $\Sigma = \{x_1 = 0\}$. See the chapter 1 for a complete description of this space.

With this notion of symbols we can characterize the piecewise smooth regularity by a transmission property of symbols. Indeed we have the well-known following result :

Proposition 5.2.1 *Denoting $x = (x_1, .., x_n) = (x_1, x')$ the variable in $I\!R^n$ and $\Sigma = \{x_1 = 0\}$, pC^{∞}_{Σ} is the set of the distributions f of $I^{\ 1}(\Sigma)$ satisfying (5.5) with*

$$a(x, \xi) \sim \sum_{j \geq 1} \frac{a_j(x')}{\xi_1^j} \text{ for } |\xi_1| > 1. \tag{5.6}$$

In this case $a_j(x') = \frac{1}{2\pi i^j} [\partial_{x_1}^j f]_{\Sigma}$ where $[\ ]_{\Sigma}$ designates the jump through Σ.

We write $pC^{\infty}_k(\Sigma) = pC^{\infty}_{\Sigma} \cap I^{-k}(\Sigma)$ for $k \in I\!N^*$.

One recalls that the asymptotic development (5.6) means for all $N \in I\!N^*$,

$$a(x, \xi) - \sum_{j=1}^{N-1} a_j(x') \frac{\chi(\xi_1)}{\xi_1^j} \in I^{-N}(\Sigma)$$

where χ is a C^{∞} function vanishing near 0 and such that $\chi(\xi_1) = 1$ if $|\xi_1|$ is large enough.

Symbols with two variables of frequency. One calls $\tilde{S}^{\mu_1,\mu_2}$ the set of the symbols with two variables of frequency defined as the C^∞ functions $b(x,\xi_1,\xi_2)$ with $x\in I\!R^n$ and $\xi_1,\xi_2\in I\!R$, so that for $\alpha\in I\!N^n$, $\beta_1,\beta_2\in I\!N$, and $\mathcal{K}$ compact of $I\!R^n$,

$$|\partial_x^\alpha\partial_{\xi_1}^{\beta_1}\partial_{\xi_2}^{\beta_2}b(x,\xi_1,\xi_2)|\le C(1+|\xi_1|)^{\mu_1-\beta_1}(1+|\xi_2|)^{\mu_2-\beta_2}$$

for all $(x,\xi_1,\xi_2)\in\mathcal{K}\times I\!R^2$.

For $\mu_1,\mu_2\in I\!R$, we defined J^{μ_1,μ_2} in the chapter 3 (see definition 3.3.1), which is the space of the ditributions f on $I\!R^n$ smooth out of $\Sigma_1=\{x_1=0\}$ and $\Sigma_2=\{x_2=0\}$ such that $f\in I^{\mu_i}(\Sigma_i)$ near each point of $\Sigma_i\backslash\Gamma$ for $i=1,2$ where $\Gamma=\Sigma_1\cap\Sigma_2$, and

$$f(x)=\int e^{i(x_1\xi_1+x_2\xi_2)}b(x,\xi_1,\xi_2)d\xi_1d\xi_2,\quad b\in\tilde{S}^{\mu_1,\mu_2}\tag{5.7}$$

near each point of the edge Γ.

We write $pC^\infty_{\Sigma_1\cup\Sigma_1}$ the space of the piecewise smooth functions with respect to $\Sigma_1\cup\Sigma_2$. Locally, their restriction to each quarter space defined by $\Sigma_1\cup\Sigma_2$ extends in a C^∞ function defined on the whole space.

In this way we have a result analogous to the proposition 5.2.1 :

Proposition 5.2.2 *We denote $x=(x_1,..,x_n)=(x_1,x_2,x'')$ the variable in $I\!R^n$, $\Sigma_i=\{x_i=0\}$ for $i=1,2$.*

Near each point of the edge $\Gamma=\{x_1=x_2=0\}$, the functions of $pC^\infty_{\Sigma_1\cup\Sigma_1}$ are the distributions $f\in J^{-1,-1}$ satisfying (5.7) with

$$b(x,\xi_1,\xi_2)\sim\sum_{j_1,j_2\ge1}\frac{b_{j_1,j_2}(x'')}{\xi_1^{j_1}\xi_2^{j_2}}\ \text{for } |\xi_1|,|\xi_2|>1.\tag{5.8}$$

In this case $b_{j_1,j_2}(x'')=\frac{1}{4\pi^2 i^{j_1+j_2}}[\partial_{x_1}^{j_1}\partial_{x_2}^{j_2}f]_\Gamma$ where $[\ ]_\Gamma$ designates the jump through Γ, that is $[f]_\Gamma=f(0^+,0^+,x'')-f(0^+,0^-,x'')+f(0^-,0^+,x'')-f(0^-,0^-,x'')$.

Let's specify the meaning of the asymptotic development (5.8) : for all $P,N\in I\!N$,

$$\begin{aligned}b(x,\xi_1,\xi_2)-&\sum_{\substack{1\le j_1\le P-1\\1\le j_2\le N-1}}b_{j_1,j_2}(x'')\frac{\chi(\xi_1)\chi(\xi_2)}{\xi_1^{j_1}\xi_2^{j_2}}\\&\equiv\sum_{1\le j_1\le P-1}b^N_{j_1}(x'',\xi_2)\frac{\chi(\xi_1)}{\xi_1^{j_1}}+\sum_{1\le j_2\le N-1}{}^Pb_{j_2}(x'',\xi_1)\frac{\chi(\xi_2)}{\xi_1^{j_2}}\end{aligned}$$

modulo $\tilde{S}^{-P,-N}$ with $b^N_{j_1} \in S^{-N}(I\!R^{n-2} \times I\!R)$ and ${}^P b_{j_2} \in S^{-P}(I\!R^{n-2} \times I\!R)$.

For $k, k' \geq 1$ we say that the symbol b verifying (5.8) is k, k'-classical if $b_{j_1,j_2} = 0$ for $j_1 \leq k$ or $j_2 \leq k'$, that is $b \in \tilde{S}^{-k,-k'}$.

Let's remark that the asymptotic development does not depend on x_1 in (5.6) and x_1, x_2 in (5.8). In fact we know (see L. Hörmander [Hö2]) that every distributions of $I^\mu(\Sigma)$ can be written under the form (5.5) with a symbol a not depending on x_1. We have an analogous result of reduction of amplitude for the elements of J^{μ_1,μ_2} :

Proposition 5.2.3 *Working near a point of the edge Γ, a distribution $f \in \tilde{S}^{\mu_1,\mu_2}$, $\mu_1, \mu_2 \in I\!R$, verifying (5.8) can be written in the form*

$$f(x) = \int e^{i(x_1\xi_1 + x_2\xi_2)} c(x'', \xi_1, \xi_2) d\xi_1 d\xi_2, \quad c \in \tilde{S}^{\mu_1,\mu_2}$$

with

$$c \sim \sum_{\alpha_1,\alpha_2 \in I\!N} \frac{1}{\alpha_1!\alpha_1!} D^{\alpha_1}_{x_1} D^{\alpha_2}_{x_2} \partial^{\alpha_1}_{\xi_1} \partial^{\alpha_2}_{\xi_2} b_{|x_1=x_2=0} \tag{5.9}$$

Here, the asymptotic development means, for all $N \in I\!N^*$,

$$c - \sum_{|(\alpha_1,\alpha_2)| \leq N-1} \frac{1}{\alpha_1!\alpha_1!} D^{\alpha_1}_{x_1} D^{\alpha_2}_{x_2} \partial^{\alpha_1}_{\xi_1} \partial^{\alpha_2}_{\xi_2} b_{|x_1=x_2=0} \in \sum_{k+k'=N} \tilde{S}^{\mu_1-k,\mu_2-k'}.$$

Moreover, let's remark that a function f in $pC^\infty_k(\Sigma_1)$ (resp. $pC^\infty_k(\Sigma_2)$) can be written under the form (5.7) where b is a k, ∞-classical symbol (resp. a ∞, k-classical symbol) that is b satisfies (5.8) with b_{j_1,j_2} and b_{j_2} (resp. b_{j_1}) vanishing for all $j_1, j_2 \in I\!N$.

Action of the pseudo-differential operators. We write $\xi = (\xi_1, .., \xi_n)$ the cotangent variables of $I\!R^n$. We need a new space of symbols : $\tilde{S}^{\mu_1,\mu_2,\nu} = \tilde{S}^{\mu_1+\nu,\mu_2} \cap \tilde{S}^{\mu_1,\mu_2+\nu}$ for $\mu_1, \mu_2, \nu \in I\!R$, corresponding to a new space of conormal distributions : $\tilde{J}^{\mu_1,\mu_2,\nu} = J^{\mu_1+\nu,\mu_2} \cap J^{\mu_1,\mu_2+\nu}$.

Let's describe the action of a pseudo-differential operator on the conormal distributions (see L. Hörmander [Hö2] for an idea of the proof).

Proposition 5.2.4 *Let P be a proper pseudo-differential operator of degree $d \leq 0$. Let's write $p(x, \xi)$ its symbol.*

i) If $f \in I^\mu(\Sigma)$, $\mu \in I\!R$, is locally given by (5.5), then $Pf \in I^{\mu+d}(\Sigma)$ can be locally written

$$Pf = \int e^{ix_1\xi_1} c(x, \xi_1) d\xi_1$$

with

$$c(x,\xi_1) \sim \sum_\alpha \frac{1}{\alpha!}\partial_\xi^\alpha p(x,\xi_1,0,..,0)D_x^\alpha a(x,\xi_1). \tag{5.10}$$

ii) If $f \in J^{\mu_1,\mu_2}$, $\mu_1,\mu_2 \in I\!R$, is locally given by (5.7), then $Pf \in \bar{J}^{\mu_1,\mu_2,d}$ can be locally written

$$Pf = \int e^{i(x_1\xi_1+x_2\xi_2)}c(x,\xi_1,\xi_2)d\xi_1 d\xi_2$$

with

$$c(x,\xi_1,\xi_2) \sim \sum_\alpha \frac{1}{\alpha!}\partial_\xi^\alpha p(x,\xi_1,\xi_2,0,..,0)D_x^\alpha b(x,\xi_1,\xi_2). \tag{5.11}$$

As defined in L. Hörmander [Hö2], the asymptotic development (5.10) means that for all $N \in I\!N^*$,

$$c(x,\xi_1) - \sum_{|\alpha|\leq N-1} \frac{1}{\alpha!}\partial_\xi^\alpha p(x,\xi_1,0,..,0)D_x^\alpha a(x,\xi_1) \in I^{\mu+d-N}(\Sigma).$$

The meaning of the asymptotic development (5.11) is the following: for all $N \in I\!N^*$,

$$c(x,\xi_1,\xi_2) - \sum_{|\alpha|\leq N-1} \frac{1}{\alpha!}\partial_\xi^\alpha p(x,\xi_1,\xi_2,0,..,0)D_x^\alpha b(x,\xi_1,\xi_2) \in \bar{J}^{\mu_1,\mu_2,d-N}.$$

In that way, if P is a parametrix of a partial differential system elliptic on a neighborhood of $N(\Gamma)$, then its action on the functions of $pC^\infty_{\Sigma_1\cup\Sigma_1}$ yields the apparition of a new type of symbols. These symbols are asymptotic sums of terms $\sigma_{j_1,j_2,j} = b_{j_1,j_2,j}(x)\frac{\rho_j(x,\xi_1,\xi_2)}{\xi_1^{j_1}\xi_2^{j_2}}$ where ρ_j is a rational function in ξ_1,ξ_2 of degree $-j \leq 0$, without any real pole, smooth with respect to x, and $b_{j_1,j_2,j}$ is a smooth function.

More precisely,

Definition 5.2.1 *For $k,k',l \in I\!N^*$, we say that a symbol $b \in \bar{S}^{-k,-k',-l}$ is k,k',l-prelogarithmic if*

$$b \sim \sum_{\substack{k\leq j_1\\ k'\leq j_2\\ l\leq j}} b_{j_1,j_2,j}(x)\frac{\rho_j(x,\xi_1,\xi_2)}{\xi_1^{j_1}\xi_2^{j_2}} \text{ for } |\xi_1|,|\xi_2|>1. \tag{5.12}$$

The asymptotic development (5.12) means : for all $P, N \in \mathbb{N}^*$,

$$b(x,\xi_1,\xi_2) - \sum_{\substack{k\le j_1\le P-1\\ k'\le j_2\le N-1\\ l\le j\le P+N-j_1-j_2-1}} \rho_{j_1,j_2,j}(x,\xi_1,\xi_2)\frac{\chi(\xi_1)\chi(\xi_2)}{\xi_1^{j_1}\xi_2^{j_2}}$$

$$\equiv \sum_{\substack{k\le j_1\le P-1\\ l\le j\le P-j_1-1}} b^N_{j_1}(x,\xi_2)\rho_j(x,\xi_1,\xi_2)\frac{\chi(\xi_1)}{\xi_1^{j_1}}$$

$$+ \sum_{\substack{k'\le j_2\le N-1\\ l\le j\le N-j_2-1}} {}^P b_{j_2}(x,\xi_1)\rho_j(x,\xi_1,\xi_2)\frac{\chi(\xi_2)}{\xi_1^{j_2}}$$

modulo $\tilde{S}^{-P,-N}$ with $b^N_{j_1} \in S^{-N}(\mathbb{R}^n \times \mathbb{R})$ and ${}^P b_{j_2} \in S^{-P}(\mathbb{R}^n \times \mathbb{R})$.

If b is such a symbol, the distribution f given by (5.7) is $pC^\infty_{\Sigma_j}$ near each point of $\Sigma_j\backslash\Gamma$ for $j = 1, 2$. Indeed, near a point of $\Sigma_1\backslash\Gamma$ for example, the variable x_2 does not vanish so we can use Taylor's development of b at 0 with respect to ξ_2 in order to make f under the form (5.5). This Taylor's development yields a symbol like (5.6).

Nevertheless, f is not $pC^\infty_{\Sigma_1\cup\Sigma_1}$ near the edge Γ in general. For example, if $b = \frac{\chi(\xi_1)\chi(\xi_2)}{\xi_1\xi_2(\xi_1^2+\xi_2^2)}$, then we have $\Delta f \equiv 4\pi^2\mathcal{H}(x_1)\mathcal{H}(x_2)$ modulo C^∞, so $\Delta\frac{\partial^2 f}{\partial x_1\partial x_2} \equiv 4\pi^2\delta$ and $\frac{\partial^2 f}{\partial x_1\partial x_2} \equiv \pi\ln(x_1^2 + x_2^2)$ modulo C^∞.

Multiplication We know that the spaces $I^\mu(\Sigma)$ ($\mu < -1$), J^{μ_1,μ_2} ($\mu_1, \mu_2 < -1$), and $pC^\infty_{\Sigma_1\cup\Sigma_1}$ are algebras. So the piecewise smooth regularity is preserved by multiplication. We are going to see that this is not available for the new type of conormal singularities described by the prelogarithmic symbols.

First, let's remind that the product of two functions written under the form (5.7) with two symbols a and b yields a similar function whose the symbol is the convolution in ξ between a and b denoted $a *_\xi b$. So the convolution between two k, k'-classical symbols gives a k, k'-classical symbol.

The particular case following is essential in order to treat our example and establishes that the product of a piecewise smooth function and an other one described by a prelogarithmic symbol can originates a new type of conormal singularity.

Let's consider $u_1 = \int e^{ix_1\xi_1} \frac{\chi(\xi_1)}{\xi_1^{k_1}}\, d\xi_1$ in $pC^\infty_{\Sigma_1}$ and $u_2 = \int e^{i(x_1\xi_1+x_2\xi_2)} \frac{\chi(\xi_1)\chi(\xi_2)}{\xi_1^{k_1}\xi_2^{k_2-1}(\xi_1-\alpha\xi_2)(\xi_1-\bar{\alpha}\xi_2)}\, d\xi_2$ with $Im(\alpha) > 0$. The symbol of u_2 is $k_1, k_2-1, 2$-prelogarithmic.

So the product u_1u_2 can be written under the form (5.7) with the symbol

$$\sigma(\xi_1,\xi_2) = \frac{\chi(\xi_1)}{\xi_1^{k_1}} *_{\xi_1} \frac{\chi(\xi_1)\chi(\xi_2)}{\xi_1^{k_1}\xi_2^{k_2-1}(\xi_1-\alpha\xi_2)(\xi_1-\bar{\alpha}\xi_2)}. \tag{5.13}$$

We have the

Proposition 5.2.5 *Let's write* $g(z) = \left[(\xi_1 - z)^{k_1} z^{k_1} \xi_2^{k_2-1}(z-\alpha\xi_2)(z-\bar{\alpha}\xi_2)\right]^{-1}$
Then the convolution defined by (5.13) is equivalent to

$$\sigma(\xi_1,\xi_2) = 2i\pi \left| \begin{array}{l} Res(g,\alpha\xi_2) \text{ if } \xi_2 > 0 \\ Res(g,\bar{\alpha}\xi_2) \text{ if } \xi_2 < 0 \end{array} \right.$$

modulo a $k_1, k_2 - 1, 2$*-prelogarithmic symbol plus a* $k_1, k_2 + 1$*-classical symbol.*

The notation $Res(g, z_0)$ *designates the residu of* g *at the pole* z_0.

Let's assume provisionally this result. We obtain :

$$\sigma(\xi_1,\xi_2) = 2i\pi \left| \begin{array}{ll} \dfrac{\chi(\xi_2)}{(\xi_1-\alpha\xi_2)^{k_1}\alpha^{k_1}(\alpha-\bar{\alpha})\xi_2^{k_1+k_2}} & \text{if } \xi_2 > 0 \\ \dfrac{\chi(\xi_2)}{(\xi_1-\bar{\alpha}\xi_2)^{k_1}\bar{\alpha}^{k_1}(\bar{\alpha}-\alpha)\xi_2^{k_1+k_2}} & \text{if } \xi_2 < 0. \end{array} \right.$$

From the form of this new symbol, we define a new class of symbols which have an asymptotic development of terms

$$\sigma_{j_1,j_2,j} = b_{j_1,j_2,j}(x) \frac{\rho_{j,sgn(\xi_2)}(x,\xi_1,\xi_2)}{\xi_1^{j_1}\xi_2^{j_2}}$$

where $sgn(\xi_2) = \left| \begin{array}{l} 1 \text{ if } \xi_2 > 0 \\ -1 \text{ if } \xi_2 < 0 \end{array} \right.$, $b_{j_1,j_2,j}$ is a smooth function, and $\rho_{j,sgn(\xi_2)}$ is, for $sgn(\xi_2)$ fixed, a rational function with respect to ξ_1, ξ_2, of degree $-j \leq 0$, without any real pole, and smooth with respect to x.

Definition 5.2.2 *For* $k, k', l \in I\!N^*$, *we say that a symbol* $b \in \bar{S}^{-k,-k',-l}$ *is* k, k', l*-logarithmic if*

$$b \sim \sum_{\substack{k\leq j_1 \\ k'\leq j_2 \\ l\leq j}} b_{j_1,j_2,j}(x) \frac{\rho_{j,sgn(\xi_2)}(x,\xi_1,\xi_2)}{\xi_1^{j_1}\xi_2^{j_2}} \quad \textit{for } |\xi_1|, |\xi_2| > 1. \tag{5.14}$$

The meaning of this new asymptotic development is similar to the one in (5.12), just replacing ρ_j by $\rho_{j,sgn(\xi_2)}$.

These symbols describe new conormal singularities. Their kind is piecewise smooth near each point of $\Sigma_1\backslash\Gamma$, but can be "logarithmic" near the points of Σ_2. For example let's consider the logarithmic symbol

$$b(x,\xi_1,\xi_2) = \frac{\chi(\xi_1)}{\xi_1^{k_1}} sgn(\xi_2)\frac{\chi(\xi_2)}{\xi_2^{k_2}}$$

The calculus of the Fourier's transformation given by (5.7) yields

$$f(x) = (c_1 x_1^{k_1-1}\mathcal{H}(x_1) + g_1)(c_2 x_2^{k_2-1} ln|x_2| + g_2)$$

where c_1, c_2 are non vanishing contants and g_1, g_2 are C^∞ functions.

Of course we could more generally consider logarithmic singularities with respect to Σ_1 too, by using $sgn(\xi_1)$. But this is not necessary to prove the Theorem 5.1.1.

The next proposition is just a generalization of the Proposition 5.2.5. It is sufficient to treat our example. One uses the generic notations $\sigma_{cl}^{k,k'}$, $\sigma_{prelog}^{k,k',l}$ and $\sigma_{log}^{k,k',l}$ which respectively designate k,k'-classical, k,k',l-prelogarithmic and k,k',l-logarithmic symbols.

Proposition 5.2.6 *If the symbol σ is k,k'-classical then*

$$\frac{\chi(\xi_1)}{\xi_1^{k_1}} *_{\xi_1} \sigma = \sigma_{cl}^{inf(k,k_1),k'}.$$

If the symbol σ is k,k',l-prelogarithmic then

$$\frac{\chi(\xi_1)}{\xi_1^{k_1}} *_{\xi_1} \sigma = \sigma_{cl}^{k_1,k'+l} + \sigma_{prelog}^{k,k',l} + \sigma_{log}^{0,k+k'+l-1,k_1}.$$

If the symbol σ is k,k',l-logarithmic then

$$\frac{\chi(\xi_1)}{\xi_1^{k_1}} *_{\xi_1} \sigma = \sigma_{log}^{k_1,k'+l,0} + \sigma_{log}^{k,k',l} + \sigma_{log}^{0,k+k'+l-1,k_1}.$$

We only demonstrate the proposition 5.2.5. See $[La]$ for more precisions.

PROOF OF THE PROPOSITION 5.2.5

Our aim is to calculate $I_R = \int_R \chi(\xi_1 - z)\chi(z)g(z)dz$.

Let $\mathcal{K} = [-R, R]$ be a compact neighborhood of 0 in $\mathbb{R}$ so that $\chi = 1$ out of it. At first we study $I_{\mathcal{K}} = \int_{\mathcal{K}} \chi(\xi_1 - z)\chi(z)g(z)dz$. Taylor's development yields, for all $N \in \mathbb{N}^*$,

$$[(z - \alpha\xi_2)(z - \bar{\alpha}\xi_2)]^{-1} = \sum_{j=0}^{N-1} z^j cst_j \frac{\chi(\xi_2)}{\xi_2^{j+2}} + z^N R_N^1(z, \xi_2). \tag{5.15}$$

So we obtain terms of the form $\int_{\mathcal{K}} \frac{\chi(\xi_1 - z)\chi(z)}{(\xi_1 - z)^{k_1} z^{k_1 - j}} dz \ \frac{\chi(\xi_2)}{\xi_2^{k_2+1+j}}$ which is a sum of the type

$$\sum_{j'=0}^{P-1} \int_{\mathcal{K}} cst_{j'} \frac{\chi(z)}{z^{k_1-j-j'}} dz \ \frac{\chi(\xi_1)\chi(\xi_2)}{\xi_1^{k_1+j'} \xi_2^{k_2+1+j}} + {}^P b(\xi_1) \frac{\chi(\xi_2)}{\xi_2^{k_2+1+j}}, \quad {}^P b \in S^{-k_1-P}(\mathbb{R}) \tag{5.16}$$

for all $P \in \mathbb{N}^*$, by using Taylor's development of $\frac{\chi(\xi_1 - z)}{(\xi_1 - z)^{k_1}}$.

Moreover the development (5.15) yields a rest

$$\int_{\mathcal{K}} \frac{\chi(\xi_1 - z)\chi(z)}{(\xi_1 - z)^{k_1} z^{k_1 - N}} R_N^1(z, \xi_2) dz \ \frac{\chi(\xi_2)}{\xi_2^{k_2+1+N}} \tag{5.17}$$

which becomes a sum of terms of the form

$$b^N(\xi_1) \frac{\chi(\xi_2)}{\xi_2^{k_2+1+j}}, \quad b^N \in S^{-k_2-1-N}(\mathbb{R}) \tag{5.18}$$

plus a rest in $\tilde{S}^{-k_1-P, -k_2-1-N}$, thanks to Taylor's development of $\frac{\chi(\xi_1 - z)}{(\xi_1 - z)^{k_1}}$.

Finally $I_{\mathcal{K}}$ is a sum of expressions of the type (5.16) and (5.18) modulo $\tilde{S}^{-k_1-P, -k_2-1-N}$ for all $P, N \in \mathbb{N}^*$, hence $I_{\mathcal{K}}$ is a $k_1, k_2 + 1$-classical symbol.

Let's treat $I_{\mathcal{K}(\xi_1)} = \int_{\xi_1 - z \in \mathcal{K}} \chi(\xi_1 - z)\chi(z)g(z)dz = \int_{\mathcal{K}} \chi(z)\chi(\xi_1 - z)g(\xi_1 - z)dz$. We carry out the study of $I_{\mathcal{K}}$ replacing the development (5.15) by

$$[(\xi_1 - z - \alpha\xi_2)(\xi_1 - z - \bar{\alpha}\xi_2)]^{-1} = \sum_{j=0}^{N'-1} z^j \sum_{l_1 + l_2 = j} cst_{l_1, l_2} (\xi_1 - \alpha\xi_2)^{-1-l_1} (\xi_1 - \bar{\alpha}\xi_2)^{-1-l_2} + z^{N'} R_{N'}^2(z, \xi_1, \xi_2) \tag{5.19}$$

The sum (5.16) is replaced by

$$\sum_{j'=0}^{P-1} cst_{j'} \frac{\chi(\xi_1)\chi(\xi_2)}{\xi_1^{k_1+j'} \xi_2^{k_2-1} (\xi_1 - \alpha\xi_2)^{1+l_1} (\xi_1 - \bar{\alpha}\xi_2)^{1+l_2}}$$
$$+ {}^P b(\xi_1) \frac{\chi(\xi_2)}{\xi_2^{k_2-1} (\xi_1 - \alpha\xi_2)^{1+l_1} (\xi_1 - \bar{\alpha}\xi_2)^{1+l_2}}, \quad {}^P b \in S^{-k_1-P}(\mathbb{R}).$$

Moreover, by noting that for all $z \in \mathcal{K}$ and $t \in [0,1]$ we have $|\xi_1 - tz - \alpha\xi_2| > cst(|\xi_1| + |\xi_2|)$ for $|(\xi_1, \xi_2)|$ large enough, we prove that the expression analogous to (5.17) defined by

$$\int_{\mathcal{K}} \frac{\chi(\xi_1 - z)\chi(z)}{(\xi_1 - z)^{k_1} z^{k_1 - N'}} R^2_{N'}(z, \xi_1, \xi_2) dz \, \frac{\chi(\xi_2)}{\xi_2^{k_2-1}}$$

is a symbol of $\bar{S}^{-k_1-P,-k_2+1-N,-2}$ if $P + N = N'$.

Hence $I_{\mathcal{K}(\xi_1)}$ is a $k_1, k_2 - 1, 2$-prelogarithmic symbol.

Finally we proved that $I_R \equiv \int_{\mathbb{R}\setminus(\mathcal{K}\cup\mathcal{K}(\xi_1))} g(z)dz$ modulo the classical and prelogarithmic symbols. The function g is analytic on $\mathbb{C}\setminus\{0, \xi_1, \alpha\xi_2, \bar{\alpha}\xi_2\}$ and can be integrated in the complex path.

As for $I_{\mathcal{K}}$ and $I_{\mathcal{K}(\xi_1)}$, $I_\gamma = \int_\gamma g(z)dz$ is still a k_1, k_2+1-classical symbol and $I_{\gamma(\xi_1)} = \int_{\xi_1 - z \in \gamma} g(z)dz$ a $k_1, k_2 - 1, 2$-prelogarithmic one, when γ is the half circle defined by $z = Re^{i\theta}$, $\theta \in [-\pi, \pi]$, and $\gamma(\xi_1) = \xi_1 + \gamma$.

Accordingly, for a tall enough $R > 0$, I_R is equivalent to $I_{\gamma_R} = \int_{\gamma_R} g(z)dz$ modulo classical and prelogarithmic symbols, where γ_R is the closed path defined by $[-R, R] \cup \gamma$.

The theorem of residus completes the proof.

Integration across Σ_1. One considers a function f satisfying (5.7) with a symbol b which can be classical, prelogarithmic or logarithmic. We are interested in the kind of the conormal singularities of $F = \int_a^{x_1} f(s, x_1, x_2)ds$ on a neighborhood of Γ, with $a < 0$.

Let's write F using b :

$$\begin{aligned} F &= \int_a^{x_1} \int e^{i(s\xi_1 + x_2\xi_2)} b(\xi_1, \xi_2) d\xi_1 d\xi_2 ds \\ &= \int e^{i(x_1\xi_1 + x_2\xi_2)} \frac{\chi(\xi_1) b(\xi_1, \xi_2)}{i\xi_1} d\xi_1 d\xi_2 \\ &\quad - \int e^{i(a\xi_1 + x_2\xi_2)} \frac{\chi(\xi_1) b(\xi_1, \xi_2)}{i\xi_1} d\xi_1 d\xi_2 \\ &\quad\quad + \int e^{ix_2\xi_2} \int \frac{e^{ix_1\xi_1} - e^{ia\xi_1}}{i\xi_1} (1 - \chi(\xi_1)) b(\xi_1, \xi_2) d\xi_1 d\xi_2. \end{aligned}$$

The first term yields a symbol $B_1 = \frac{\chi(\xi_1)b(\xi_1,\xi_2)}{i\xi_1}$. The second one corresponds to the function described by B_1 restricted to the plane $\{x_1 = a\}$ which intersects transversely Σ_2 outside Γ. So the conormal singularities with respect to Σ_1 disappears and one deduces the kind of the symbol B_2 associated to the second term. For example if b is k, k', l-prelogarithmic, then B_2 is $\infty, k'+l, 0$-prelogarithmic.

It remains to be study the third term. Noting that $\frac{e^{ix_1\xi_1}-e^{ia\xi_1}}{i\xi_1}(1-\chi(\xi_1))$ is C^∞ with a compact support, a Taylor's development of b at 0 with respect to the variable ξ_1 implies there is not any conormal singularity along Σ_1. So the type of the symbol B_3 describing the last term is the same as for B_2.

In that way we obtain the form of the symbol $B = B_1 + B_2 + B_3$ of F :

Proposition 5.2.7 *Let f be under the form (5.7) with a symbol b. One denotes B the symbol of $F = \int_a^{x_1} f(s, x_1, x_2)ds$ with $a < 0$.*

Then $B = \sigma_{cl}^{k+1,k'}$ *if b is k, k'-classical,*

$B = \sigma_{prelog}^{k+1,k',l} + \sigma_{cl}^{\infty,k'+l}$ *if b is k, k', l-prelogarithmic,*

$B = \sigma_{log}^{k+1,k',l} + \sigma_{log}^{\infty,k'+l,0}$ *if b is k, k', l-logarithmic.*

5.3 Application: appearance of logarithmic singularities

From the regularity of the solution espressed by (5.4), we are going to compute the complete symbol of v describing its conormal singularities with respect to $\Sigma_2^+ = \{x_2 = 0, x_1 > 0\}$. The idea is to define a map $\mathcal{T}$ for which v is invariant. In this way we will approach v by a sequence of functions built by iteration with $\mathcal{T}$.

First let's recall that u is known and the partial system (5.2) is microlocally elliptic on a neighborhood of $N(\Gamma)$. So using a microlocal parametrix of (5.2), and working near 0 so that $\psi(x_1 + x_2 + x_3) = 1$, we obtain

$$w_1 \equiv P(uv) \text{ modulo } C^\infty$$

where P is a pseudo-differential operator of order -1 with the symbol

$$\tau(\xi_1, \xi_2, \xi_3)\frac{\xi_2}{i[\xi_2(\xi_2 - 2\xi_3) + (\xi_1 - \xi_3)^2]}$$

where the function τ is equal to 1 on a neighborhood of $N(\Gamma)$ and vanishes outside the domain of ellipticity.

As $\partial_{x_1} v = uw_1$, we have

$$v(x_1, x_2, x_3) = \int_a^{x_1} uP(uv)(s, x_2, x_3)ds + v(a, x_2, x_3).$$

We fix $a < 0$ so that the term $v(a,.,.)$ is in $I^{-k_2}(\Sigma_2)$ if we work locally near 0, by hypothesis in the past.

Finally we can define the map

$$\mathcal{T} : w \mapsto \int_a^{x_1} uP(uw)(s, x_2, x_3)ds + v(a, x_2, x_3),$$

so that $v \equiv \mathcal{T}(v)$ modulo C^∞.

Let's built the sequence $(v_k)_k$ by iteration :

$$\begin{aligned} v_0 &= v(a,.,.) \in I^{-k_2}(\Sigma_2) \\ v_{k+1} &= \mathcal{T}(v_k) \text{ for } k \in \mathbb{N}, \end{aligned}$$

which allows us to state precisely the conormal singularities of v with respect to $\Sigma_2 \backslash \Gamma$.

Proposition 5.3.1 *For $k \in \mathbb{N}$, $v_k - v$ is an element of $J^{-k_1-1,-k_2-k}$.*

PROOF

We know that $v_0 - v$ is an element of $J^{-k_1-1,-k_2}$.
Let's assume that $v_k - v \in J^{-k_1-1,-k_2-k}$ for some $k \in \mathbb{N}$.

As $u \in I^{-k_1}(\Sigma_1)$ we have $u(v_k - v) \in J^{-k_1,-k_2-k}$ and applying the pseudo-differential operator P of degree -1 we obtain $P[u(v_k - v)] \in J^{-k_1,-k_2-k-1}$, so $uP[u(v_k - v)] \in J^{-k_1,-k_2-k-1}$.
Finally $v_k - v \equiv \mathcal{T}(v_k) - \mathcal{T}(v)$
$\equiv \int_a^{x_1} uP[u(v_k - v)]ds$ modulo C^∞
is in $J^{-k_1-1,-k_2-k-1}$.
In that way we just need to compute by iteration the symbol of each v_k to express the asymptotic development of the symbol describing the conormal singularities of v with respect to $\Sigma_2 \backslash \Gamma$.

For $k \in \mathbb{N}$ we denote σ_k the complete symbol of v_k, that is we can write $v_k = \int e^{i(x_1\xi_1+x_2\xi_2)}\sigma_k(x,\xi_1,\xi_2)d\xi_1 d\xi_2$ locally.

Since v is piecewise smooth with respect to Σ_2 in the past, v_0 is in $pC^{\infty}_{k_2}(\Sigma_2)$, and we can say that σ_0 is a ∞, k_2-classical symbol.

Let's compute σ_1. From the fact that $v_0 \equiv \int e^{ix_2\xi_2} c_2 \frac{\chi(\xi_2)}{\xi_2^{k_2}} d\xi_2$ modulo C^∞ and $u \equiv \int e^{ix_1\xi_1} c_1 \frac{\chi(\xi_1)}{\xi_1^{k_1}} d\xi_1$ modulo C^∞, we deduce the form of the complete symbol of uv_0 :

$$\sigma_{uv_0} = c_1 c_2 \frac{\chi(\xi_1)\chi(\xi_2)}{\xi_1^{k_1}\xi_2^{k_2}} + \sigma_{cl}^{\infty,k_2}(\xi_1,\xi_2) + \sigma_{cl}^{k_1,\infty}(\xi_1,\xi_2).$$

Using the proposition 5.2.4 we obtain the complete symbol of $P(uv_0)$:

$$\begin{aligned}\sigma_{P(uv_0)} &\equiv \frac{\xi_2}{i(\xi_1^2+\xi_2^2)}\sigma_{uv_0} \text{ modulo } C^\infty \\ &= \frac{c_1c_2}{i}\frac{\chi(\xi_1)\chi(\xi_2)}{\xi_1^{k_1}\xi_2^{k_2-1}(\xi_1-i\xi_2)(\xi_1+i\xi_2)} + \sigma_{cl}^{\infty,k_2+1} + \sigma_{cl}^{k_1+1,\infty}.\end{aligned}$$

The multiplication between u and $P(uv_0)$ yields the convolution between their symbols, and the symbol of $uP(uv_0)$ is

$$\begin{aligned}\sigma_{uP(uv_0)} = c_1^2 c_2 \frac{\chi(\xi_1)}{\xi_1^{k_1}} *_{\xi_1} \frac{\chi(\xi_1)\chi(\xi_2)}{\xi_1^{k_1}\xi_2^{k_2-1}(\xi_1-i\xi_2)(\xi_1+i\xi_2)} \\ + \frac{\chi(\xi_1)}{\xi_1^{k_1}} *_{\xi_1} \sigma_{cl}^{\infty,k_2+1} + \frac{\chi(\xi_1)}{\xi_1^{k_1}} *_{\xi_1} \sigma_{cl}^{k_1+1,\infty}\end{aligned}$$

Thanks to the proposition 5.2.6,

$$\sigma_{uP(uv_0)} = \sigma_{log} + \sigma_{prelog}^{k_1,k_2-1,2} + \sigma_{cl}^{k_1,k_2+1}$$

and the proposition 5.2.5 gives

$$\sigma_{log}(\xi_1,\xi_2) = \pi c_1^2 c_2 \left| \begin{array}{ll} \dfrac{\chi(\xi_2)}{i^{k_1+1}(\xi_1-i\xi_2)^{k_1}\xi_2^{k_1+k_2}} & \text{if } \xi_2>0 \\ \dfrac{\chi(\xi_2)}{(-i)^{k_1+1}(\xi_1+i\xi_2)^{k_1}\xi_2^{k_1+k_2}} & \text{if } \xi_2<0. \end{array}\right.$$

By integration, see the proposition 5.2.7, we obtain the symbol of $v_1 - v_0 = \int_a^{x_1} uP(uv_0)ds$:

$$\sigma_{v_1-v_0} = \sigma_{princ\,log} + \sigma_{prelog}^{k_1+1,k_2-1,2} + \sigma_{cl}^{k_1+1,k_2+1}$$

with
$$\sigma_{princ\,log} = \sigma_{log}^{1,k_1+k_2,k_1} + \sigma_{log}^{\infty,2k_1+k_2,0}.$$

More precisely, noticing that $\int e^{i(x_1\xi_1+x_2\xi_2)}\sigma_{princ\,log}(\xi_1,\xi_2)d\xi_1 d\xi_2$ is in the space $I^{-k_1,-k_1-k_2}(\Sigma_1,\Gamma)$, the property $(P_{4.11}), a)$ implies

$$\int e^{ix_1\xi_1}\sigma_{princ\,log}(\xi_1,\xi_2)d\xi_1 \equiv \sigma_{princ\,log}(0,\xi_2) \text{ modulo } S^{-\infty}(I\!R)$$

near each point of Σ_2^+.

Let's remark this is a symbol in $S^{-2k_1-k_2}(I\!R)$ which does not verify the transmission property characterizing the symbols of the piecewise smooth functions.

So we have proved the following result:

Proposition 5.3.2 $\sigma_1 = \sigma_0 + \sigma_{cl}^{k_1+1,k_2+1} + \sigma_{prelog}^{k_1+1,k_2-1,2} + \sigma_{princ\,log}$ *where*

i) $\sigma_{princ\,log} = \sigma_{log}^{1,k_1+k_2,k_1} + \sigma_{log}^{\infty,2k_1+k_2,0}$,

ii) $\int e^{ix_1\xi_1}\sigma_{princ\,log}(\xi_1,\xi_2)d\xi_1 \equiv c\,sgn(\xi_2)\frac{\xi_2}{\xi_2^{2k_1+k_2}}$ *with* $c \neq 0$.

Iterating this process, it is no more difficult to demonstrate the

Proposition 5.3.3 *For all* $k \in I\!N^*$, *we have*

$$\sigma_k \equiv \sigma_1 + \sigma_{cl}^{k_1+1,k_2+1} + \sigma_{prelog}^{k_1+2,k_2-1,2} + \sigma_{prelog}^{k_1+1,k_2,2} + \sigma_{log}^{1,k_1+k_2+1,k_1}$$

modulo $\tilde{S}^{-k_1-1,-2k_1-k_2-1}$.

Because the classical and prelogarithmic symbols describe piecewise smooth singularities near Σ_2^+, any other type of singularity for v_k originates in the logarithmic symbols. Nevertheless all of them are in $\tilde{S}^{-k_1-1,-2k_1-k_2-1}$ except $\sigma_{princ\,log}$ which corresponds to the principal no piecewise smooth singularity with respect to Σ_2^+.

So we have

$$v_k(x) \equiv \int e^{ix_2\xi_2}\sigma_{princ\,log}(0,\xi_2)d\xi_2 \text{ modulo } pC^{\infty}_{k_1+k_2+1}(\Sigma_2)+I^{-2k_1-k_2-1}(\Sigma_2)$$

near each point of Σ_2^+.

The calculus of the Fourier's transform yields near Σ_2^+ :

$$v_k(x) \equiv c x_2^{2k_1+k_2-1} ln|x_2| \text{ modulo } pC^{\infty}_{k_1+k_2+1}(\Sigma_2) + I^{-2k_1-k_2-1}(\Sigma_2)$$

with $c \neq 0$.

The Proposition 5.3.1 allows us to conclude.

Principal notations

We generally use the basic notations of the theory of Partial Differential Operators (see L. Hörmander [Hö]):

$D_{x_j} = \frac{1}{i}\frac{\partial}{\partial x_j}$, $D^\alpha = D_{x_1}^{\alpha_1}..D_{x_n}^{\alpha_n}$ if $\alpha = (\alpha_1, .., \alpha_n) \in \mathbb{N}^n$,
$|\alpha| = \alpha_1 + .. + \alpha_n$.
$C^\infty(\Omega)$: Space of smooth functions in Ω
$C_0^\infty(\Omega)$: Space of smooth functions in Ω with compact support
$\mathcal{D}'(\Omega)$: Space of distributions in Ω
S^μ : Space of global symbols of degree μ
$\Psi^\mu(\Omega)$: Space of proper pseudodifferential operators of degree μ in Ω.
$\Psi_{cl}^\mu(\Omega)$: Space of classical proper pseudodifferential operators of degree μ in Ω: $p(x,\xi) \sim \sum_{j\in\mathbb{N}} p_j(x,\xi)$ with $p_j(x,\xi)$ homogeneous of degree $\mu - j$ with respect to ξ, $p =$ complete symbol.
OpS^μ : Space of global pseudodifferential operators
$a(x, D) = Opa$: Global pseudodifferential operators of complete symbol $a(x,\xi)$.
$H_s(\mathbb{R}^n), H_s(\Omega)$: Sobolev spaces of index s
${}^\infty H_s(\mathbb{R}^n), {}^\infty H_s(\Omega)$: Besov spaces of index s
$WF_s(u)$: Wave front set of u relatively to H_s^{loc} or ${}^\infty H_s^{loc}$
$WF(u)$: Wave front set of u relatively to C^∞.

We use also the following notations:

$T(X)$: Tangent bundle to X
$T^*(X)$: Cotangent bundle to X
$N(\Sigma) \subset T^*(X)$: Conormal bundle to $\Sigma \subset X$
pC_Σ^∞ : Piecewise smooth functions with respect to Σ
$H_{s,k}(\Sigma)$: k-Conormal distributions with respect to Σ, based on H_s^{loc} or ${}^\infty H_s^{loc}$ (see Definition 1.2.1), $H_s(\Sigma) = H_{s,\infty}(\Sigma)$.

Bibliography

[A] Alinhac S. Interaction d'ondes simples pour des équations complètement non linèaires. Ann. Scient. Ec. Norm. Sup., 21, p. 91-132, 1988.

[AG] Alinhac S., Gerard P. Opérateurs pseudo-differentiels et Théorème de Nash-Moser. InterEditions/Editions du CNRS, 1991.

[Be] Beals M. [1] Spreading of singularities for a semilinear wave equation. Duke Math. J., 49, p. 275-286, 1982.
————-[2] Self spreading and strengh of singularities for solutions to semilinear waves equations. Annals of Math., 118, p. 187-214, 1983.
————-[3] Propagation of smoothness for nonlinear second order strictly hyperbolic equations. Proc. of Symp. in Pure Math., 43, p. 21-24, 1985.
————-[4] Propagation and Interaction of singularities in nonlinear hyperbolic problems. Birkhaeuser, Boston, 1989.

[BM] Beals M., Metivier G. Reflection of transversal progressing waves in nonlinear strictly hyperbolic mixed problems. American journal of math. , 109, p.335-360, 1987.

[Bo] Bony J.M. [1] Interaction des singularités pour les équations aux dérivées partielles non linéaires. Sém. Goulaouic-Schwartz, Ecole Polytechnique, 22, Paris, 1979-1980.
————-[2] Interaction des singularités pour les équations aux dérivées partielles non linéaires. Sém. Goulaouic-Schwartz, 2, Paris, 1980-1981. Ecole Polytechnique.
————-[3] Propagation et Interaction des singularités pour les équations aux dérivées partielles non linéaires. Proc. Int. Cong. Math., p.1133-1146, Warsawas, 1983.

[Bou] Bougrini H. Interaction des symboles principaux pour les ondes conormales semi-linéaires monodimensionnelles. C.R.A.S., Paris, p.1205-1210, 1995.

[BPV] Bougrini H., Piriou A., Varenne J.P. Propagation et Interaction pour les ondes conormales semi-linéaires . Commun. in Partial Differential Equations, 23, p.333-370, 1998.

[CP] Chazarain J., Piriou A. Introduction à la théorie des équations aux dérivées partielles. Gauthier-Villars, Paris, 1981.

[Ch] Chemin J.Y. Calcul paradifférentiel précisé et applications à des équations aux dérivées partielles non semi linéaires. Duke Math. J., 56, p. 431-469, 1988.

[De] Delort J.M. Conormalité des ondes semi linéaires le long des caustiques. Amer. J. Math., 113, p. 593-631, 1991.

[D] Duistermaat J.J. Fourier Integral Opertors. Progress in Math., Birkhäuser, 1996.

[DH] Duistermaat J.J., Hörmander L. Fourier Integral Opertors II. Acta Math. 128, p.183-269, 1972.

[GU] Greenleaf A., Uhlmann G. Estimates for singular Radon transforms and pseudodifferentiel operators with singular symbols. Jour. Funct. Anal., 89, p.202-232, 1980.

[Hö] Hörmander L. [1] Fourier Integral Operators I. Acta Math., 127, p.79-183, 1971.

————-[2] The analysis of partial differential operators. Springer-Verlag, Berlin, I, II, III, IV, 1983-1985.

————-[3] Lectures on nonlinear hyperbolic differential equations. Berlin, Heidelberg, Paris, 1997.

[J] Joshi M. S. A commutator proof of the propagation of polyhomogeneity for semi-linear equations. commun. in P.D.E. , 22, (3-4), p.435-463, 1997 .

[La] Laschon G. Singularités précisées pour les ondes conormales semi-linéaires. Thèse, univ de Nice, 1998.

[Lax] Lax P.D. Asymptotic solutions of oscillatory initial value problems. Duke Math. J., 24, p. 627-646, 1957.

[Le] [1] Lebeau G. Problème de cauchy semi-linéaire en 3 dimensions d'espace. Un résultat de finitude, Jour. Func. Anal. 78, p.185-196. 1988.

————-[2] Equations des ondes semi-linéaires II. Contrôle des singularitès et cuastiques semi-linéaires. Invent. Math, 95, p. 277-323, 1989.

[MRi] Melrose R.B., Ritter N. Interaction of progressing waves through a nonlinear potential. In Sém. Goulaouic-Shwartz, 12, Paris, p.1983-84. Ecole Polytechnique.

[MU] Melrose R.B., Uhlmann G. [1] Lagrangian intersection and the Cauchy problem. Comm. in Pure and Appl. Math., p. 483-519, 1979.

————-[2] Interaction of nonlinear progressing waves for semilinear wave equations. Ann. of Math., 121, p.187-213, 1985.

————-[3] Interaction of progressing waves for semilinear wave equations II. Ark. Mat., 25, p. 91-114, 1987.

[M] Métivier G. [1] The Cauchy problem for semilinear hyperbolic systems with dicontinuous data , Duke Math. J. , 53, p.983-1011, 1986.

————-[2] Propagation, interaction, and reflection of discontiuous progressing waves. Amer. J. Math., 111, p.239-289, 1989.

[MR] Métivier G. and Rauch J. [1] The interaction of two progressing waves. Springer Lecture Notes in Math., Springer-Verlag, New York, 1402, p.216-226, 1989.

————-[2] Interaction of piecewise smooth progressing waves for semilinear hyperbolic equations. Comm. In P.D.E., 15, p.239-289, 1990.

[Mi] Mizohata S. Systèmes hyperboliques. J. Math. Soc. Japan 11, p.205-233, 1959.

[NP] Nadir B., Piriou A. [1] Symboles pour deux ondes conormales sans interaction. C.R.A.S, 305, Paris, série I, p.337-340, 1987.

————-[2] Ondes semi-linéaires conormales par rapport à deux hypersurfaces transverses. Duke Math. J., 58, p.577-599, 1989.

[NV] Nadir B., Varenne J.P. [1] Problème de Cauchy et reflexion transeverse pour une onde conormale classique. C.R.A.S., Paris, t.309, serie I, p.817-820, 1989.

————-[2] Régularité conormale classique des problèmes de Cauchy et de réflexion transverse pour un système 2×2 semi-linéaire. Ann. Inst. Fourier, tom 40, fascicule 4, p.849-866, 1990.

[P] Piriou A. Calcul symbolique non linéaire pour une onde conormale simple. Ann. Inst. Fourier, 28, p.173-187, 1988.

[PV] Piriou A., Varenne J.P. [1] Ondes semi-linéaires discontinues conormales par morceaux. C.R.A.S., Paris, tome 311, série I, p.607-610, 1990.

————[2] Problème de Cauchy semi-linéaire à donnée initiale conormale: régularité à l'arête et symbole principal. J. Math. Pures et Appl., 75, p.9-49, 1996.

[RR] Rauch J., Reed M. [1] Jump discontinuities of semilinear strictly hyperbolic systems in two variables : creation and propagation. Comm. Math. Phys., 81, p.203-227,1981.

————[2] Nonlinear microlocal analysis of semilinear hyperbolic problems in one space dimension. Duke Math. J., 49, p.397-475, 1982.

————[3] Discontinuous progressing waves for semilinear systems. Comm. In P.D.E., 10, p.1033-1075,1985.

————[4] Classical conormal solutions of semilinear systems. Comm. In P.D.E., 13, p.1297-1335, 1988.

————[5] Classical conormal semilinear waves, Séminaire Equations aux dérivées partielles, Ecole Polytechnique, exposé 5, 1985-86.

[R] Ritter N. Progressing waves solutions to nonlinear hyperbolic Cauchy problems, Ph.D. Thesis, M.I.T., 1984.

[S] Sà Barreto A. [1] Interactions of conormal waves for fully semilinear wave equations. J. Fonct. Anal., 89, p. 233-273, 1990.

————-[2] On the interactions of conormal waves for semilinear wave equations, in Microlocal Analysis and Nonlinear Waves. IMA Vol Math. Appl., 30, Springer Verlag, New York, p. 1-17, 1991.

————-[3] Second microlocal ellipticity and propagation of conormality for semilinear wave equations. J. Fonct. Anal., 102, p. 47-71, 1991.

[V] Varenne J.P. Etude d'ondes conormales semi-linéaires: structure précisée à l'arête, développements classiques . Thèse d'Etat, Univ. de Nice, 1993.

[W] Wang Weike, Interaction of conormal waves with different singularities for semilinear equations. Journal of P.D.E. (China), 4, 1991.

Index